2D Semiconductors for Environmental Remediation

This book gives a comprehensive description of various aspects of 2D semiconductors including their synthesis, surface science, characterizations, and their allied application in environmental remediation including air and water purification, oil-water separation, hydrogen production, and CO_2 removal. The electronic and optoelectronic enhancement properties of these semiconductors with bandgap engineering, doping, and chemical functionalization for various applications are exemplified.

Features:

- Provides focus on the application of 2D semiconductors for environmental applications.
- Covers the fundamental understanding of material design, fabrication, defect engineering, physical and chemical properties of 2D semiconductors, and their advantages for environmental applications.
- Focuses on the reliable future perspectives and developments that lead to advanced research in utilizing 2D materials for large-scale exploration and commercial deployment.
- Explores an in-depth insight into various aspects and processes for environmental remediation.
- Details the importance of 2D semiconductors over other 0D, 1D, or 3D semiconductors.

This book is aimed at researchers and graduate students in materials science and environmental engineering.

Emerging Materials and Technologies

Series Editor: Boris I. Kharissov

The *Emerging Materials and Technologies* series is devoted to highlighting publications centered on emerging advanced materials and novel technologies. Attention is paid to those newly discovered or applied materials with potential to solve pressing societal problems and improve quality of life, corresponding to environmental protection, medicine, communications, energy, transportation, advanced manufacturing, and related areas.

The series takes into account that, under present strong demands for energy, material, and cost savings, as well as heavy contamination problems and worldwide pandemic conditions, the area of emerging materials and related scalable technologies is a highly interdisciplinary field, with the need for researchers, professionals, and academics across the spectrum of engineering and technological disciplines. The main objective of this book series is to attract more attention to these materials and technologies and invite conversation among the international R&D community.

Friction Stir-Spot Welding
Metallurgical, Mechanical and Tribological Properties
Edited by Jeyaprakash Natarajan and K. Anton Savio Lewise

Phase Change Materials for Thermal Energy Management and Storage
Fundamentals and Applications
Edited by Hafiz Muhammad Ali

Nanofluids
Fundamentals, Applications, and Challenges
Shriram S. Sonawane and Parag P. Thakur

MXenes
From Research to Emerging Applications
Edited by Subhendu Chakroborty

Biodegradable Polymers, Blends and Biocomposites
Trends and Applications
Edited by A. Arun, Kunyu Zhang, Sudhakar Muniyasamy and Rathinam Raja

Bioinspired Materials and Metamaterials
A New Look at the Materials Science
Edward Bormashenko

Computational Studies
From Molecules to Materials
Edited by Ambrish Kumar Srivastava

2D Semiconductors for Environmental Remediation
*Edited by Honey John, Nisha T Padmanabhan, Sona Stanly,
and Jith C Janardhanan*

For more information about this series, please visit: www.routledge.com/
Emerging-Materials-and-Technologies/book-series/CRCEMT

2D Semiconductors for Environmental Remediation

Edited by
Honey John, Nisha T. Padmanabhan,
Sona Stanly, and Jith C. Janardhanan

CRC Press
Taylor & Francis Group
Boca Raton London New York

CRC Press is an imprint of the
Taylor & Francis Group, an **informa** business

Designed cover image: www.shutterstock.com

First edition published 2025
by CRC Press
2385 NW Executive Center Drive, Suite 320, Boca Raton FL 33431

and by CRC Press
4 Park Square, Milton Park, Abingdon, Oxon, OX14 4RN

CRC Press is an imprint of Taylor & Francis Group, LLC

ISBN: 9781032381916 (hbk)
ISBN: 9781032381923 (pbk)
ISBN: 9781003343899 (ebk)

DOI: 10.1201/9781003343899

Typeset in Times
by codeMantra

Contents

Preface

The editors are delighted to present the book entitled "*2D Semiconductors for Environmental Remediation*" in the high-quality research content of the book series "Emerging Materials and Technologies." The book title was so-chosen as it converges the importance and application of two-dimensional (2D) nanomaterials in environmental remediation. The book can be a significant breakthrough in bringing together researchers, academicians, and graduate/postgraduate students from science and engineering backgrounds interested in 2D nanomaterials and their allied applications for a virtuous environment. The book comprehensively describes the state-of-the-art aspects of 2D semiconductors, including their synthesis, surface science, characterizations, and allied application in environmental remediation. Various fields, like air and water purification, oil-water separation, hydrogen production, CO_2 removal, etc., are explained in detail. The electronic and optoelectronic enhancement properties of these semiconductors with bandgap engineering, doping, and chemical functionalization for various applications are illustrated. The book also covers the most recent advances in fabrication, a deeper understanding of the principles underlying their engineering of electronic and optical behavior to allied applications, mechanistic insights, and an evaluation of their effectiveness and perspectives for environmental management. Equal emphasis is given to both material features and application aspects of 2D semiconductors.

Two-dimensional semiconductors have attracted the scientific community since its first report regarding the exfoliation of an elemental layered semiconductor in 2014, while graphene monolayer struck the research community earlier in 2004. The reactivity, pressure, optoelectronic properties, and electric field can be tuned with the number of layers that directly influence the anisotropic properties of these 2D materials, which also have gained tremendous research interest. The area of 2D semiconductors always had rapid progress, with each time added with a new member added to the family. But still, the study of 2D nanomaterials and their commercialization is in its infancy. In short, the basic idea of this book is to provide an overview of the recently developed 2D semiconductors and to stress their recital in environmental remediation. A total of eight chapters have been accepted for publication in this book, which is supposed to furnish key insights and fundamental concepts for experts, early career researchers, and students of material science, chemical engineering, and environmental engineering.

Chapter 1 gives a fine introduction to 2D nanomaterials; concentrates on the early innovations that led to the discovery of 2D materials; classifications based on structure and composition; and descriptively reveals the nature, physical, and chemical properties of different 2D nanomaterials. In addition, the nature of escalating interest in this area and the causes thereof are reported. The surface science of specific 2D nanomaterials is also exemplified.

Chapter 2 briefs on the advanced synthesis techniques and structural design of various 2D semiconductors and their heterostructures. The different physical and chemical approaches for the preparation of graphene, MXenes, and transition metal chalcogenides, categorized as top-down and bottom-up approaches, their critical applications, and their limitations are discussed. The ease of tunability to attain unique physical and chemical properties by chemical functionalization is also added.

Chapter 3 defines the basic principle of photocatalytic hydrogen production, the cutting-edge 2D semiconductor materials and its heterostructures employed, and their recent development protocols. The importance of engineering the electronic, surface, structural, and optical properties of these 2D semiconductors for enhanced hydrogen production is also explained in detail, together with current difficulties and potential solutions.

Chapter 4 highlights the utilization of 2D semiconductor materials for the selective adsorption of heavy metals and dyes from industrial effluents, which would otherwise interfere with natural water resources and cause severe toxicological issues to the environment and human health. The adsorption properties of 2D-layered materials such as MoS_2, graphene oxide, MXenes, metal-organic frameworks, MMT nanoclay, layered double hydroxide for both organic dyes, and heavy metal ions are neatly typified.

Chapter 5 includes the photocatalytic mechanism of CO_2 reduction and the advantages of utilizing photocatalytic 2D nanomaterials and their heterostructures for CO_2 photoreduction in terms of the extended absorption spectrum, enhanced carrier separation efficiency, improved electronic behavior, etc. Photocatalysis is a promising solution for both global warming and the energy crisis, which could transform the greenhouse gas CO_2 into solar chemicals and fuels. Despite the fact that there is now a lot of research being done in this field, many advances have to be brought out to realize commercial and industrial applications.

Chapter 6 discusses the recent developments in nanotechnology and 2D nanomaterials in removing air contaminants and minimizing their effects on human health. The methodology of purifying air includes selective adsorption strategies and photocatalytic processes. A particular focus is given to the semiconductor TiO_2 nanoparticles and 2D nanosheets of graphene, g-C_3N_4, demonstrating substantial performance toward the decomposition of volatile organic compounds. These air-purifying technologies are more environmentally friendly than traditional ones to remove air pollutants.

Chapter 7 describes the photocatalytic membranes based on 2D nanomaterials for water purification processes, including the photodegradation of organic pollutants, photoinduced antimicrobial treatment, development of photocatalytic membranes and aerogels, etc. Particular emphasis is given to the state-of-the-art g-C_3N_4 hybrids and their immobilization to photocatalytic membranes for water purification technologies and environmental remediation and concludes with potential difficulties in commercializing.

Chapter 8 summarizes the broad scope, including various commercialized application fields and products based on 2D semiconductor nanomaterials. Finally, a unique perspective of the ongoing research with future challenges and outlooks of 2D semiconductors in the field of environment applications is briefed.

The perspectives discussed in different sessions in the chapters, as mentioned above, amply demonstrate the need for multiple pilot-scale initiatives to be made in the upcoming decade to commercialize 2D nanomaterials successfully for the environment prosperity. Finally, we would like to sincerely thank each author for their significant contributions to the publication of this book. We gratefully applaud the editorial and publishing staff of the Taylor and Francis Group for their excellent assistance and support.

Honey John
Nisha T. Padmanabhan
Sona Stanly
Jith C. Janardhanan

Editors

Honey John obtained her Ph.D. from Cochin University of Science and Technology (Kerala, India) in 2004. She joined the Indian Institute of Space Science and Technology (IIST), Trivandrum, in 2007, as a reader and worked to the grade of Associate Professor. In 2015, she re-joined the Department of PS&RT, CUSAT as a professor. In 2017, she became the honorary Director of Inter University Centre for Nanomaterials and Devices, CUSAT. Her current research areas are triboelectric and piezoelectric nanogenerators, 2D nanoscrolls with magnetic nanoparticles, photo-catalytic self-cleaning materials, hydrogen production, aerogels for supercapacitors and pollutant removal, magnetoplasmonics, and green tires. Prof. Honey has a strong research background in Materials Science, Chemical Engineering, and Polymer Technology, for over 18 years. Her research team has recently innovated novel materials as energy-harvesting triboelectric nanogenerators, alongside developing a novel wet chemical approach to synthesize 2D nanoscrolls. Her research interests also critically span over the development of magnetic hybrids, supercapacitors from functional materials, semiconductors for photocatalytic/electrocatalytic hydrogen production and CO_2 reduction to value-added fuels, self-cleaning materials and coat-ings, conducting polymers, development of green tires, synthesis of polymeric aero-gels for pollutant removal, and organic/inorganic nanohybrids for optical/electrical/ sensor applications. She has one granted patent (three filed and under evaluation), 65 high-impact refereed journal publications, 15 international book chapters, and more than 70 international/national conference papers to her credit. She has received about ₹1.65 Cr. research grants to support her work. Currently, she is also the prin-cipal coordinator of CUSAT for the advanced research in "Tactical Oceanography, Modern Sensors, Signal Processing and Telemetry"—a funded project (₹1.83 Cr.) by DRDO, India. She is a reviewer of many international journals, including ACS, RSC, Elsevier, Wiley, and Taylor and Francis. She has signed MoU with different autonomous bodies of Govt. of Kerala and many polymer industries. Prof. Honey is a resource person for many international/national conferences, refresher courses, and expert member/chairperson of many Ph.D. evaluation committees of many central/ Kerala universities, university exam/interview boards, etc. She holds many adminis-trative responsibilities in CUSAT and is a Board of Studies Member of many colleges and the departments in CUSAT.

Nisha T. Padmanabhan, Assistant Professor, Department of Chemistry and Centre for Research, St. Teresa's College (Autonomous), Ernakulam, pursued her doctorate in 2021 from Cochin University of Science and Technology (CUSAT), Kerala, India. She earned her Masters in Chemistry from Mahatma Gandhi University in 2012 and Bachelors in Education in 2014. Her interests are on developing 2D nanomaterials, especially graphene, borophene, MoS_2, and MXenes, for energy- and environment-related applications. She is focused to develop, evaluate, and demonstrate the performance of cost-effective electrodes or/and photoelectrodes for electrocatalytic/photoelectrocatalytic water splitting for hydrogen production and conversion of CO_2 to desired chemical feedstocks, along with the development of photocatalytic membranes and aerogels for water purification, crystal/phase engineering, and electrospun nanofibers for triboelectric nanogeneration.

Jith C. Janardhanan obtained bachelor and master degrees in Chemistry from the University of Calicut, India. He received Ph.D. in Chemistry (Organic Chemistry) at Cochin University of Science and Technology, India in 2019. After a short stay at CSIR-NIIST, he joined CUSAT as a postdoctoral researcher. His current research interest focus on the development of novel methods for the synthesis of small organic molecules and heterocycles, development of color-tunable fluorescent organic dyes and its applications, development of 2D-layered materials for supercapacitor, hydrogen evolution, and heavy metal adsorption studies.

Sona Stanly pursued her doctoral degree in Polymer Technology from Cochin University of Science and Technology in 2021. She was an ad-hoc faculty member in the Department of Polymer Science and Rubber Technology, CUSAT, from 2010 to 2017 and joined the Department of Technical Education, Government of Kerala in 2019 as a lecturer in Polymer Technology. She is presently Assistant Professor at Govt. Polytechnic College, Koratty, Thrissur. Her research interests are in 2D materials, especially graphene, clays, hybrids, and polymer composites for environmental remediation.

Contributors

Irthasa Aazem
Nanotechnology and Bio-Engineering
Research Group, Department of
Environmental Science,
Health and Biomedical (HEAL)
Strategic Research Centre,
Atlantic Technological University,
ATU Sligo, Ash Lane, Sligo, Ireland

A. Aishu
Department of Environmental Studies,
Kannur University,
Kannur, Kerala, India

V. P. Animasree
Department of Applied Chemistry
Cochin University of Science and
Technology
Kochi, Kerala, India

K. P. Anjali
Department of Chemical Engineering,
National Institute of Technology
Durgapur,
Durgapur, West Bengal, India

S. Anjitha
Department of Environmental Studies,
Kannur University,
Mangattuparamba Campus
Kannur, Kerala, India

M. J. Avani
Department of Applied Chemistry,
Cochin University of Science and
Technology,
Kochi, Kerala, India

Lim Ying Chin
School of Chemistry and Environment,
Faculty of Applied Sciences,
Universiti Teknologi MARA,
Shah Alam Selangor, Malaysia

Priyanka Ganguly
Chemical and Pharmaceutical Sciences,
School of Human Sciences,
London Metropolitan University,
London, United Kingdom

Suja Haridas
Department of Applied Chemistry,
Inter-University Centre for
Nanomaterials and Devices,
Cochin University of Science and
Technology,
Kochi, Kerala, India

Jith C. Janardhanan
Department of Polymer Science and
Rubber Technology,
Cochin University of Science and
Technology,
Kochi, Kerala, India

P. J. Jandas
Department of Polymer Science and
Rubber Technology,
Cochin University of Science and
Technology,
Kochi, Kerala, India
Department of Chemistry,
CoE, Sensors and Nanoelectronics,
CMR Institute of Technology,
Marathahalli, Bengaluru, India

E. J. Jelmy
Department of Polymer Science and
Rubber Technology,
Cochin University of Science and
Technology,
Kochi, Kerala, India

Merin Joseph
Department of Applied Chemistry,
Cochin University of Science and
Technology,
Kochi, Kerala, India

Devagi Kanakaraju
Chemistry Programme,
Faculty of Resource Science and
Technology,
Universiti Malaysia Sarawak,
Sarawak, Malaysia

Vishal Kandathil
Department of Applied Chemistry,
Cochin University of Science and
Technology,
Kochi, Kerala, India

M. K. Kavitha
U R Rao Satellite Centre, Indian Space
Research Organisation,
Bengaluru, Karnataka, India

K. Manoj
Department of Environmental Studies,
Kannur University,
Mangattuparamba Campus,
Kannur, Kerala, India

Narayanapillai Manoj
Department of Applied Chemistry,
Inter University Centre for
Nanomaterials and Devices,
Cochin University of Science and
Technology,
Kochi, Kerala, India

Michael A. Morris
School of Chemistry and Advanced
Materials and Bioengineering
Research Centre (AMBER),
Trinity College Dublin,
Dublin, Ireland

Nisha T. Padmanabhan
Inter University Centre for
Nanomaterials and Devices,
Cochin University of Science and
Technology,
Ernakulam, Kerala, India

Sibu C. Padmanabhan
School of Chemistry and Advanced
Materials and Bioengineering
Research Centre (AMBER),
Trinity College Dublin,
Dublin, Ireland

P. Periyat
Department of Environmental Studies,
Kannur University,
Mangattuparamba Campus,
Kannur, Kerala, India

Suresh C. Pillai
Nanotechnology and Bio-Engineering
Research Group, Department of
Environmental Science,
Health and Biomedical (HEAL)
Strategic Research Centre,
Atlantic Technological University, ATU
Sligo,
Ash Lane, Sligo, Ireland

Sebastian Nybin Remello
Department of Applied Chemistry,
Inter-University Centre for
Nanomaterials and Devices,
Cochin University of Science and
Technology,
Kochi, Kerala, India

C. V. Sijla Rosely
Department of Polymer Science and
    Rubber Technology,
Cochin University of Science and
    Technology,
Kochi, Kerala, India

Sona Stanly
Govt: Polytechnic College,
Koratty, Thrissur, Kerala, India
Department of Technical Education,
Government of Kerala, India

Rinku Mariam Thomas
Department of Mathematics and
    Natural Sciences,
School of Arts and Science,
American University of Ras Al
    Khaimah,
United Arab Emirates

1 The Surface Science of 2D Semiconductors
An Introduction

M. K. Kavitha and Nisha T. Padmanabhan

1.1 INTRODUCTION

In the past two decades, a great deal of research has been done on nanomaterials to either provide brand-new solutions or improve upon those that already exist for many of the most urgent environmental issues, from water scarcity to air pollution (Wang and Mi 2017). Two-dimensional (2D) nanomaterials, from pristine graphene to graphene oxide (GO), reduced GO (rGO), and later to hexagonal boron nitride (h-BN), MXenes, graphitic carbon nitride, silicene, germanene, and borophene, etc. drawn a lot of attention due to their significant specific surface area and fascinating features that are often not obtainable in their bulk counterparts. Nanomaterials based on graphene have so far been used in a wide range of environmental applications, including membrane-based separation, photocatalytic oxidation, sensing, and adsorptive removal of impurities (Saba and Jawaid 2018). The unique properties of graphene attracted a lot of attention initially because of its 2D, sheet-like structure and characterizations. Research on other non-carbon-based materials grew as a consequence. As a result, 2D materials have recently attracted a lot of attention because they offer excellent electrical, mechanical, and optical capabilities for various real-world applications. Researchers from all around the world are becoming more interested in 2D materials due to their distinctive properties and promising futures in a variety of engineering sectors. Over the course of a decade, this eventually led to the isolation of numerous 2D materials.

This chapter introduces various 2D nanomaterials of enormous interest concerning their structure, peculiar properties, and categorizations. Investigating novel 2D materials with flexible structures and varied properties offers chances to broaden the 2D family and also to evade the drawbacks of the existing ones. Though fascinating, the existing library of 2D nanomaterials is still insufficient to meet all environmental applications with potency. Consequently, researchers modified the surface and interface of current materials to introduce remarkable topographies and thereby tuned properties. This chapter thus defines the need for surface modifications and defect engineering in 2D nanomaterials to attain specific performance and potency toward various environmental applications. Finally, the implications of 2D nanomaterials on deployment to the environment and future research needed to nurture the environmental performance are also briefed.

DOI: 10.1201/9781003343899-1

1.2 HISTORICAL DEVELOPMENT OF 2D NANOMATERIALS

Before the discovery of graphene by Novoselov and Geim in 2004 (Novoselov et al. 2004), scientists almost believed that the existence of 2D nanomaterials in nature was not conceivable due to their environmental instability. However, it is theoretically deduced that there are currently more than 700 2D nanomaterials, even though some of them have not yet been synthesized. Even though graphite was known from the 16th century and was industrially used for the production of steel, dry lubricants, and brake linings, it was after the discovery of fullerenes in 1985 (Kroto et al. 1985) and one-dimensional (1D) carbon nanotubes in 1991 (Iijima 1991), that the existence of countless new carbon allotropes became clear. Earlier in 1962, Boehm introduced the suffix "ene" for materials of a single sheet. In 1994, about ten years before the discovery of graphene, germanene, and silicene were undergone initial investigations of their electrical and atomic structures. These discoveries spurred an enormous amount of study, encouraging scientists to create 1D nanoribbons or separate 2D graphite materials. Numerous findings on fullerenes, graphite, and nanotubes have since been reported, with graphite being considered to be the original material. However, it wasn't until 2004 that an endeavor to isolate a graphene monolayer was made viable (Novoselov et al. 2004). This was the start of significant advancement for a new class of materials.

Graphene, a monolayer of graphite, can be regarded as a primitive 2D nanomaterial that can be isolated with stability in a single-atom thickness. Graphene layers were plucked and transferred to SiO_2 (thin layer) over a silicon wafer using the Scotch tape method (Novoselov et al. 2004). The so-created 2D nanomaterial was far thinner than paper and more robust than diamond, with an electrical conductivity much higher than copper. There are numerous ways to prepare the same today, thanks to graphene research. Consequently, graphene was the first contemporary carbon-based 2D nanomaterial. Its development demonstrated that stable, single-, and few-atom-thick layer materials are feasible and capable of exhibiting outstanding technologically advantageous features. Researchers began to examine and synthesize non-carbon-based 2D materials after the discovery of graphene. Since then, numerous other 2D materials have been discovered. Later, the research attention was gained by transition metal dichalcogenides (TMDs) due to their outstanding electrical properties, transparency, and elasticity and has demonstrated applications in catalysis, photovoltaics, transistors, photodetectors, Li-ion batteries, etc. (Choi et al. 2017). The sandwich structure of TMDs, which has two chalcogens sandwiching the transition layer of metal, distinguishes it from graphene's structure. TMDs, primarily including molybdenum disulfide (MoS_2), tungsten disulfide (WS_2), molybdenum diselenide ($MoSe_2$), and tungsten diselenide (WSe_2), can be regarded as the first graphene offsprings that spurred substantial research effort.

Later, in 2011, Drexel University synthesized the first MXene of the chemical form $M_{n+1}X_nT_x$ ($Ti_3C_2T_x$ being the first known MXene), by precisely etching MAX phases (Naguib et al. 2011). Within a short period, about 40 different variants of MXenes, including V_2CT_x, Ti_2CT_x, Mo_2CT_x, Nb_2CT_x, and $Ti_3C_2T_x$, have been produced and can be considered as one of the fastest-growing 2D nanomaterial families; many more have been theoretically anticipated as well (Gogotsi and Anasori 2019). Moreover, in the past three to four years, a significant number of research has been reported with MXenes. 2D layers of boron, silicon, and germanium, respectively as borophene, silicene, and germanene, along with h-BN, are the other 2D nanomaterials that were explored with much interest. The growth of silicene was first reported in 2012 on Ag (111) surfaces by molecular beam epitaxy (MBE) (Vogt et al. 2012). However, unlike their 2D structures, silicene properties starkly contrast with those of graphene, as they don't contain crystals with 3D layers in their parents. After the successful fabrication of silicene, the scientific interest moved to borophene. In 2015, three distinct borophene phases were synthesized successfully by depositing elemental boron on Ag (111) surfaces under an ultrahigh vacuum (Mannix et al. 2015). Theoretical simulations and atomic-scale characterization revealed formations resembled fused boron clusters with out-of-plane buckling (Mannix et al. 2015). Unlike boron, borophene exhibits metallic properties as found in 2D metals of high anisotropy. In the search of germanene synthesis, germanane (GeH), the fully hydrogen-terminated companion of germanene, was initially created using the topochemical deintercalation of the layered van der Waals (vdW) solid calcium digermanide (Bianco et al. 2013). After that, in 2014, researchers at Aix-Marseille University in France discovered germanene by MBE using Au (111) surfaces (Dávila et al. 2014). 2D growth of tin (Sn) as stanine was first reported by MBE (Zhu et al. 2015). Still, many more 2D materials can be explored similarly from their parent 3D compounds. Also, all these atomically thin 2D nanomaterials can be characterized as metallic, semi-metallic, semiconducting, insulating, or superconducting depending on their chemical and structural arrangements.

1.3 CLASSIFICATION OF 2D SEMICONDUCTORS

In general, 2D materials formed by the stacking of individual atomic layers are characterized by a layered crystal structure with strong in-plane bonding, and the layers are held together by weak vdW forces, resulting in unexpected physical and chemical properties (Geim and Grigorieva 2013). The critical property which determines the semiconducting nature is the bandgap. A plethora of 2D materials possess semiconducting bandgaps from microwave to ultraviolet (UV) regions in the electromagnetic spectrum. Due to the large surface-to-volume ratio, their band structure is susceptible to external perturbations. Consequently, tuning of bandgap by doping, intercalation, or even external electric field opens the scope of 2D semiconductors for a wide range of applications (Figure 1.1).

1.3.1 BASED ON COMPOSITION

1.3.1.1 Graphene-Based Materials

Pristine graphene is a single layer of carbon atoms arranged in a hexagonal honeycomb structure. It is obtained by the micromechanical cleavage of high-quality graphite; this method is not scalable for large-scale production of graphene. The scalable approach to producing high-quality graphene includes epitaxial growth on transition metals and chemical vapor deposition (CVD) using hydrocarbons (Zhang, Zhang, and Zhou 2013). Graphene exhibits exceptional properties like lightweight,

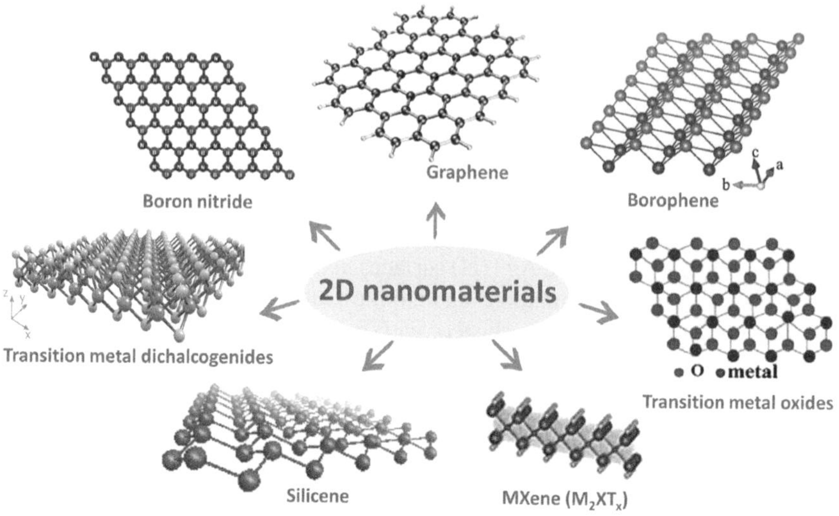

FIGURE 1.1   Structures of some 2D nanomaterials.

robust mechanical strength, huge surface area, and high thermal and electrical conductivity. Although graphene has excellent properties, its zero-bandgap and hydrophobicity limit its application in certain areas. Nowadays, functionalized graphene or graphene-derived materials are gaining attention as efficient materials for environmental applications. Graphene-based materials like GO, rGO, and its hybrids are identified as excellent adsorbents for contaminants, photocatalysts for the degradation of organic pollutants, water separation membranes, electrodes for sensing and removal of heavy metal ions (Allgayer, Yousefi, and Tufenkji 2020; Zhao, Ni, et al. 2019; Seenivasan, Chang, and Gunasekaran 2015).

1.3.1.1.1 GO

One of the most popular approaches to obtaining graphene-based materials is through GO. GO has a layered carbon structure with oxygen functional groups such as hydroxyl (OH), the carboxylic acid (COOH), ketone (C=O), epoxy (–O–) groups attached on both sides of the plane and the edges (Dreyer et al. 2010). It can be single-layered, few-layered, or multilayered structures. GO can be prepared by the oxidation of graphite followed by physical or chemical exfoliation. Since GO is highly oxidized, it acts as an electrical insulator. However, the presence of functional groups makes it hydrophilic and can form stable suspension in aqueous solutions. Thus, GO is widely utilized as the source for synthesizing graphene, significantly rGO, other graphene derivatives like fluorographene, hybrids with small organic molecules, polymers, etc. The carbon-to-oxygen (C/O) atomic ratio is an essential parameter for GO which predicts its hydrophilicity, chemical reactivity, sheet resistance, and dispersion in water and organic solvents. The C/O ratio generally varies from 1.5 to 2.5 depending on the synthesis methods (Araújo et al. 2017). In general, GO is synthesized by the oxidation of graphite. GO was first synthesized in 1859 by B. C. Brodie by the oxidation of graphite with potassium perchlorate as an oxidizing agent in the presence of fuming nitric acid. The product corresponds to a C/O ratio of 2.2, and Brodie named it graphic acid (Brodie 1859). Further, Brodie's method was modified by Staudenmaier by adjusting the chlorate and sulfuric acid into the mixture; the product obtained shows similar properties to Brodie-GO. The most accepted and efficient method is Hummers' method using excess potassium permanganate, sulfuric acid, and a small amount of sodium nitrate (Hummers and Offeman 1958). Hummers' GO also shows a very similar C/O ratio of 2.25. Since the Hummers method is not environmentally friendly due to the evolution of oxide of nitrogen during reaction, several modifications are made including nitrate free, varying sulfuric acid concentration, room/low temperature etc.

Apart from chemical methods, electrochemical synthesis of GO is considered as eco-friendly and key to large-scale production. It consists of oxidation involving graphite working electrodes and graphite counter electrodes, dipped in electrolytes. The electrochemical process involves the intercalation of cations or anions followed by oxidation. The yield of GO depends on the type of electrolyte used due to its different oxidation capacities. Acid electrolyte like H_2SO_4, $HClO_4$, inorganic salts like $(NH_4)_2SO_4$, Na_2SO_4, K_2SO_4 are used. Acid electrolytes like H_2SO_4 give GO with a C/O ratio of less than 2, indicating a greater extent of oxidation (Pei et al. 2018). Low concentrations of acid electrolytes create strain in the carbon lattice and more amorphous graphite structures. Higher concentration favors more intercalation and maintains the crystallinity of carbon lattice (Lowe et al. 2019).

1.3.1.1.2   rGO

rGO is obtained by the reduction of GO. rGO is considered as a good compromise between graphene and GO. Since rGO has structural properties close to graphene, depending on the nature of the functional groups present and is easy to prepare from GO. The reduction can be accomplished by thermal, chemical, electrochemical, photo-assisted, or microwave methods. Thermal reduction involves the annealing of GO at high temperatures in an inert or reducing atmosphere. During annealing, exfoliation and removal of oxygen functional groups via liberation of CO_2 and CO occur, resulting in the formation of graphene sheets with large lateral sizes (Zhao et al. 2010). Annealing the temperature plays a vital role in the rate of reduction; a C/O ratio was no more than 7 was obtained, if the temperature was less than 500°C while an higher C/O ratio of higher than 13 was attained when a temperature 750°C was attained (Mattevi et al. 2009; Rai, Kavitha, and Jaiswal 2015). Thermal reductions by unconventional heating methods include microwave irradiation and photo-irradiation. Voiry et al. demonstrated microwave reduction of solution-exfoliated GO. The as-obtained microwave rGO shows a highly ordered structure with high mobility, >1,000 square centimeters per volt per second (Voiry et al. 2016). Irradiating focused sunlight to reduce GO to rGO is a green and clean method for producing a large quantity of rGO (Eswaraiah, Jyothirmayee Aravind, and Ramaprabhu 2011; Yar et al. 2021).

Chemical reduction of GO can be made using different reducing agents like sodium borohydride, L-ascorbic acid, hydrazine, sodium or potassium hydroxide, hydroiodic acid, etc. Hydrazine is a good regent for the reduction of GO. The product obtained shows a C/O ratio of 10, which is highly agglomerated due to an increase in hydrophobicity. For further fabrication of chemically rGO for different applications, a colloidal state in water is maintained by adding polymers like PVP or/and PVA as surfactants (Swain and Bahadur 2014). Sodium borohydride is more effective in removing carbonyl groups in GO (Shin et al. 2009). Ascorbic acid is a mild and environmentally friendly reducing agent for GO, achieving a C/O ratio of 12.5 (Fernández-Merino et al. 2010). Moreover, the colloidal rGO obtained does not allow accumulation. Hydroiodic acid is a potent reducing agent enabling fine control over the reduction of GO, and the C/O ratio of the rGO is 15 (Pei et al. 2010).

GO is either suspended in solution or deposited on the electrode for electrochemical reduction. The reduction was performed in electrolytes like LiCl, KCl, and KNO_3 with an aqueous buffer solution. The reduction occurs by the electron exchange between GO and electrodes, and reduction can be controlled by the pH of the buffer solution (Zhou et al. 2009)

1.3.1.2 Single-Element 2D Semiconductors

Borophene (B), aluminene (Al), gallenene (Ga), indiene (In), and thallene (Tl) are some of the single-element graphene-like 2D materials which possess 2D morphology. Besides, other layered 2D materials such as silicene (Si), germanene (Ge), stanene (Sn), plumbene (Pb), phosphorene (P), arsenene (As), antimonene (Sb), bismuthene (Bi), selenene (Se), and tellurene (Te) also belong to the same family of single-element graphene-like 2D materials. They are closely related to graphene or graphene derivatives functionalized with hydrogen ions, metals, or hydroxyl groups. For mechanical cleavage, CVD or MBE is suggested for preparing these materials. Among the family of these 2D materials, group VA and VIA elements show semiconducting nature. Zhang et al. reported elemental black phosphorus 2D material with an adjustable bandgap from 0.3 to 2 eV (Castellanos-Gomez et al. 2014). Similar to black phosphorus, b-arsenic (b-As) has an orthorhombic structure with layered configuration, so that monolayer or few-layer b-As is synthesized by Zhong et al. by micromechanical cleavage (Zhong et al. 2018). Further, antimonene and bismuthene are also prepared by epitaxial methods. Group VIA 2D materials involve selinene and tellurene with semiconducting characteristics (Lozovoy et al. 2022).

1.3.1.3 *h*-BN

h-BN is promising 2D-layered material similar to graphene. Analogous to graphene, h-BN possesses similar lattice parameters and bonding orbital; they are called "white graphene", though it includes opposite electronic properties. B and N form sp^2 hybrids with hexagonal planar structure, the electronegativity difference between B and N results in confined electron movement. h-BN is an excellent semiconductor which possesses a wide bandgap along with high thermal and chemical stability. Cassabois et al. demonstrated the h-BN has an indirect bandgap value of 5.955 eV by two-photon absorption spectroscopy (Cassabois, Valvin, and Gil 2016). Due to the wide optical bandgap, h-BN shows strong absorption in the deep UV region and UV luminescent band at 215 nm attributed to band-to-band transition. Due to the wide bandgap, h-BN will act as a UV photocatalyst.

As the number of layers increases, there is a reduction in the bandgap of h-BN. The edge structure and presence of defects play an essential role in the electronic properties of h-BN. BN nanosheets show insulating nature with high resistivity (10^{18} Ω.cm) (Wang et al. 2022). However, edge functionalization with OH, NO_2, and halogens can change the bandgap of BN nanoribbons (Lopez-Bezanilla et al. 2011; Liu et al. 2018).

Different synthesis methods for h-BN include micromechanical cleavage, CVD (Koepke et al. 2016), and chemical routes. Micromechanical cleavage involves ball milling and ultrasonication. Moderate shear force is applied in ball milling to exfoliate bulk h-BN. The yield and quality of h-BN nanosheets obtained by milling depend on different parameters like ball size, milling speed, ball-to-powder ratio, and milling agent. To minimize planar structural defects and effective functionalization, appropriate milling agents are chosen. For example, ball milling in the presence of sodium hydroxide gives hydroxyl-functionalized BN nanosheets. Liquid-phase exfoliation assisted by sonication in organic solvents is found to be a simple and more promising method for the synthesis of BN nanosheets (Lin, Williams, and Connell 2010). Organic solvents like N,N-dimethylformamide, dimethyl sulfoxide, N-methyl pyrrolidone, and isopropyl alcohol can break the interlayer vdW forces via ultrasonication. The addition of suitable ionic surfactants or stabilizers increases the stability of the as-obtained dispersion. Ye et al. demonstrated that stable dispersions of high-quality monolayer or few-layer BN nanosheets can be prepared from h-BN via sonication-mediated exfoliation by employing hyperbranched polyethylene stabilizer in both tetrahydrofuran and chloroform (Ye et al. 2018). The CVD method is utilized for synthesizing large area defect free h-BN using molecular precursors like borazine, ammonia borane, or the combination of BCl_3 and NH_3 or BF_3 and NH_3 (Koepke et al. 2016; Babenko et al. 2017; Choi 1999; Jang et al. 2016; Hidalgo et al. 2013). BN nanosheets grow on transition metal substrates like Cu and Ni, Pt, Pd, etc.; the chemical synthesis method involves mixing boron precursor and nitrogen precursor and heating at high temperatures, for example, synthesizing submicron-sized BN using boric acid and melamine at 400 °C. Further, h-BN was ball milled to obtain BN nanosheets with uniform size distribution (Li et al. 2021).

Surface modification of BN nanosheets is applied to customize their properties. Covalent functionalization of h-BN with OH and NH groups shows a smaller bandgap of 4.2 eV, and further, these functionalizations will improve the photocatalytic activity compared to the pristine form (Liu et al. 2018). Moreover, functionalizing h-BN will enhance its water solubility and dispersibility (Gautam and Chelliah 2021). Non-covalent functionalization of h-BN nanostructures with visible photocatalysts like TiO_2, ZnO, metal nanoparticles has been found to be effective in improving photocatalytic activities (Wang et al. 2018). The BN nanostructures were always negatively charged intrinsically, which makes them good hole-acceptor. Thus incorporation of BN will effectively improve the separation and transfer of photogenerated charge carriers (Fu et al. 2013).

1.3.1.4 TMDs

TMD compounds have the general formula MX_2, M-transition metal, and X-chalcogen. Individual TMD monolayer comprises hexagonally packed M atoms sandwiched by two layers of X atoms. For MX_2, the M is group IV metals, generally Mo, W and the X is chalcogen atoms among which S, Se, and Te are commonly seen as in reported TMDs. Depending on the atomic stacking configurations, MX_2 can form two crystal structures: trigonal prismatic (2H) and octahedral (1T) phases. MoS_2 and WS_2 are 2D chalcogenides widely used for environmental applications. MoS_2 nanosheets have a bandgap near 1.9 eV, which allows them to act as a visible light photocatalyst (Li, Meng, and Zhang 2018). The CVD method is used widely to grow these inorganic layered materials. Typically, a substrate is coated with oxide of Mo or W by physical vapor deposition, and then these films are exposed to S or Se powder (Zhou et al. 2018). Depending upon the growth conditions and concentration of precursors, continuous films, triangular or other polygonal, are formed. Liquid-phase exfoliation involves sonication of bulk MoS_2 or WS_2 in the presence of LiOH, n-BuLi, organic solvents like NMP, and functionalized polymers (Velusamy et al. 2015; Cunningham et al. 2012). The wet chemical method is an appropriate alternative for CVD and liquid phase exfoliation. However, the chemical synthesis of TMDs still faces challenges due to the non-availability of proper reagents and inefficiency in achieving high purity and yield. MoS_2 nanosheets can be prepared using ammonium molybdate and thiourea by solvothermal method (Xie et al. 2013).

1.3.1.5 MXenes

MXenes are 2D materials of the family of carbides and nitrides of transition metals. 2D flakes of MXene have the general formula $M_{n+1}X_nT_x$ and possess an intermixed pattern of n layers of carbon or nitrogen (X) with $n+1 (n = 1-3)$ layers of transition metals (M). The T_x in the formula represents the surface terminations, such as O, OH, F, and/or Cl, which are bonded to the outer M layers. The identity and specific concentration of these surface functionalities on the MXene surface are highly dependent on the nature of synthetic methods. Thus, MXenes have the general formula $M_4X_3T_x$, $M_3X_2T_x$, and M_2XT_x. The first MXene ($Ti_3C_2T_x$) was discovered in 2011 (Naguib et al. 2011). MXenes have the unique feature that two transition metals are mixed in such a way that either as solid solution $(Ti,Nb)CT_x$ or sandwich structure $Mo_2TiC_2T_x$. MXenes are generally synthesized by etching A layer from the MAX phase using HF, HCl-NaF, or HCL-LiF. For example, $Ti_3C_2T_x$ is obtained by the etching of Ti_3AlC in HF and followed by exfoliation. The etching and exfoliation result in the enhancement of hydrophilic nature due to the functionalization of MXenes with abundant terminated hydroxyl and fluoride-terminated functional groups (Hope et al. 2016). MXenes with functionalized surfaces possess negative zeta potential and are hydrophilic to act as effective adsorbents for environmental pollutant purification and sensing of heavy metal ions. MXenes have a strong adsorption capacity for metal ions, even more potent than carbon nanotubes and GO. $Ti_3C_2(OH/ONa)_xF_{2x}$ MXene materials have reported adsorbing Pb(II) ions (140 mg/g) with high selectivity (Peng et al. 2014). $Ti_3C_2T_x$ MXene with fluorine functionalization shows a high Cr(VI) adsorption capacity of 250 mg/g under optimized synthetic conditions. Moreover, these MXenes reduce Cr(VI) to less toxic Cr(III) (Ying et al. 2015).

1.3.2 BASED ON STRUCTURES

2D materials can assemble into new nanostructures, due to their sizeable lateral size, ultrathin thickness, and high flexibility. Based on their structure, the self-assembly of

2D materials can be classified into three: 1D, 2D, and three-dimensional (3D). These structures will provide an enormous specific surface area and excellent mechanical stability. These nano-structural materials are developed from 2D semiconductors and are utilized for various applications like adsorbents, photocatalysts, membranes, sensors, etc.

1.3.2.1 1D

Ultrathin sheets of graphene, TMDs render efficient building blocks for constructing 1D nanostructures like nanotubes, nanoscrolls, nanowires, and nanofibers. Xue et al. developed mesoporous h-BN nanofibers by pyrolysis with a large surface area of ~620 m^2/g. These nanofibers show fast and selective adsorption of organic pollutants with good cyclic stability (Xue et al. 2014). Li et al. synthesized h-BN whiskers with a high surface of ~960 m^2/g, and the obtained whiskers possess parallel and layered structures. These 2D nanostructures show excellent adsorption performance due to the presence of high-density defects and micro/mesopores. These features provide active sites for adsorption. Moreover, non-covalent interaction between organic pollutants and h-BN will also provide contribution to the adsorption (Li et al. 2016). Graphene nanoscrolls are formed by the rolling of single- or a few-layered graphene sheets. Generally, this assembly is obtained by the intercalation and sonication of graphite. Viculis et al. demonstrated the preparation of carbon scrolls first by intercalating graphite followed by sonication in ethanol (Viculis, Mack, and Kaner 2003). Template-assisted assembly is an effective method to obtain 1D structures. For example, carbon nanotubes and metal oxide nanowires are used as a template for preparing graphene nanoscrolls around these tubular structures (Yan et al. 2013). Chiral nanofibers can be prepared from 2D sheets of GO, and TMDs by vigorous stirring in the presence of polymers like Pluronic P123. Votex-assisted vigorous stirring causes the twisting of nanosheets into 1D chiral nanowires and polymer will help to preserve the chirality (Tan et al. 2015).

1.3.2.2 2D

Multilayer films, coatings, or free-standing membranes can be fabricated using 2D materials as building blocks. 2D material-based membranes are extensively used for water desalination (Foller, Wang, and Joshi 2022; Liu et al. 2020). Nanosheet-based membranes perform selective mass transportation via two mechanisms, namely size exclusion and Donnan electrostatic exclusion. The size exclusion mechanism selectively permeates small water molecules through in-plane nanopores or nanochannels in between the layers while blocking the hydrated ions. In the Donnan mechanism, an electrically charged surface blocks the hydrated ions. Intercalation of 2D materials like GO with foreign molecules will modify the interlayer spacing. This helps to develop robust membranes with stable and high performance. GO laminates intercalated with water molecules show NaCl rejection up to 97% (Song et al. 2018). The stacking of GO sheets can prepare multilayered stable free-standing membranes. The highly ordered arrangement of GO sheets in the film resulted in 2D nanochannels between two adjacent graphene sheets which enabled the selective sole permeation of water without undesired solutes (Joshi et al. 2014). The presence of oxygen-containing functional groups enables related charge-based interactions with water pollutants.

Layer-by-layer (LBL) self-assembly is a powerful technique for preparing multilayer films. Generally, LBL films are prepared by the alternate coating of oppositely charged polymer layers. However, LBL method can modify for the fabrication of 2D materials composite films (Kavitha, Gopinath, and John 2015). GO can be utilized in LBL films due to the versatility in surface charge, typically negatively charged, but can be modified to positive surface charge with amine functionalization. TMDs like MoS$_2$ have a negative surface charge at neutral pH; further, it can be limited to a positive charge by functionalization with cationic polymers (Gupta, Arunachalam, and Vasudevan 2015; Hu and Yang 2021). This type of surface modification in 2D materials opens the scope for the fabrication of LBL films for different applications like nanofiltration, chemical sensors, photocatalysts, etc. (Hu and Yang 2021).

1.3.2.3　3D

3D materials can be used to construct hierarchical 3D structures such as core structures, aerogel assembly, and 3D porous structures (Chen et al. 2022; Zhuang et al. 2021; Masud, Zhou, and Aich 2021; Zhou, Yu, and Jiang 2017). 3D assemblies of 2D materials are developed by CVD, hydrothermal and solvothermal routes following either template-assisted or non-template methods. 3D porous graphene free-standing structures can be synthesized using a 3D metal scaffold for graphene growth. For example, CVD growth of graphene on a 3D nickel template gives free-standing porous sponge-like graphene (Zhou, Yu, and Jiang 2017). Rasch et al. developed a versatile wet-chemical assembly of 2D carbon nanomaterials that allows their shaping into cubic centimeter-sized, 3D macroscopic networks via a ceramic template. This method is demonstrated for the assembly of electrochemically exfoliated graphene, GO, and rGO with tailored porosity and pore size with remarkable mechanical stability. This method is proposed for assembling other 2D materials like TMDs or MXenes (Rasch et al. 2019). Maleki et al. synthesized microporous h-BN structure by pyrolysis using cetyltrimethylammonium bromide and ammonia borane (Maleki, Beitollahi, and Shokouhimehr 2015). h-BN-based porous structures are helpful for wastewater treatment, hydrogen evolution, and catalyst support.

1.4　ENGINEERING PROPERTIES OF 2D SEMICONDUCTORS FOR ENVIRONMENTAL APPLICATIONS

2D semiconducting nanomaterials exhibit fascinating chemical, mechanical, electrical, and optoelectronic properties that are crucial for various environmental applications, as well as electrodes for batteries, capacitors, and other energy storage or conversion devices. Since 2D nanomaterials have significant benefits over their bulk materials in terms of surface qualities, they are extensively researched and used for a variety of applications. Large surface-to-volume ratios and contact areas; facilitated charge carrier migration to the reaction sites on the surface due to the presence of an atomically thin layer; highly dense electronically active surface defects that facilitate the adsorption of target molecules; a large number of active sites due to the availability of surface atoms; controllable functionalities at their low dimensions; high porosity which aids in enhancing absorption; uniformity in dispersions made possible by their low-dimensional nanostructures; suitability for constructing heterostructures with the same or other low-dimensional materials such as 2D/1D or 2D/0D interfaces, and the ability to combine these to produce novel materials with emergent synergistic phenomena, etc. are some of the distinguishing qualities of 2D nanomaterials that make them particularly suitable for environmental applications.

Even while a lot of pristine 2D materials can be directly applied for various environmental applications like water treatment, air purification, hydrogen production, etc., their efficacy in energy conversions is too meager to be used/deployed in commercial mode. These 2D semiconductor materials have to be thus engineered in order to increase their conversion efficiencies either by bandgap tuning or interface engineering (Figure 1.2). Various methodologies are adopted to modify the bandgap of 2D materials, including thickness reduction, surface functionalization, and dopant inclusion. Additionally, artificial 2D/2D, 2D/1D, or 2D/0D nanostructures could be constructed for increased activity compared to their single 2D counterparts by forming heterostructures of materials of different low dimensions. Surface modifications and interface/defect engineering can significantly impact bandgap states, surface dipole, energy level alignment, orbital interactions, etc., thus being advantageous to environmental applications, including utilizing renewable solar energy. As atoms for substitutions, modifications by covalent or non-covalent interactions, are available on the surface of 2D semiconductors, tuning their physical and electronic properties is incredibly achievable. Additionally, artificial 2D/2D nanostructures could be constructed for increased activity compared to their bulk counterparts by forming heterostructures of different 2D materials. Following are the engineering approaches adopted in the case of 2D materials to achieve better environmental applications.

1.4.1 Surface Chemical Functionalization

Primarily, surface modification of materials has a significant impact on catalytic performances since catalysis is mostly a surface-related phenomenon. Surface termination of 2D semiconductors with various metal or non-metal functional groups by coordination facilitates a large surface area. As a result, anchoring of different functional groups on these nanostructured surfaces can be easily accomplished using straightforward chemical processing techniques. These methods enable the attachment of diverse functional groups to the surface, allowing for the modulation of surface architecture and surface charges to meet specific requirements. This flexibility offers chances to design and fine-tune unique features at the interface while also enabling control over the surface composition.

Surface functionalization can be covalent or non-covalent in nature. The molecule with which the 2D nanomaterial surface is functionalized can be electron donors or acceptors, such that they can change the local charge state of the 2D nanomaterial, leading to substantial changes in the electronic properties. Non-covalent interactions are typically vdW forces, hydrogen bonding, electrostatic interactions, or coordination bonds. For instance, aligned or non-aligned CNTs and graphene functionalized/adsorbed with poly(diallyldimethylammonium chloride) (PDDA) can perform as catalysts for metal-free oxygen reduction reaction (ORR) in fuel

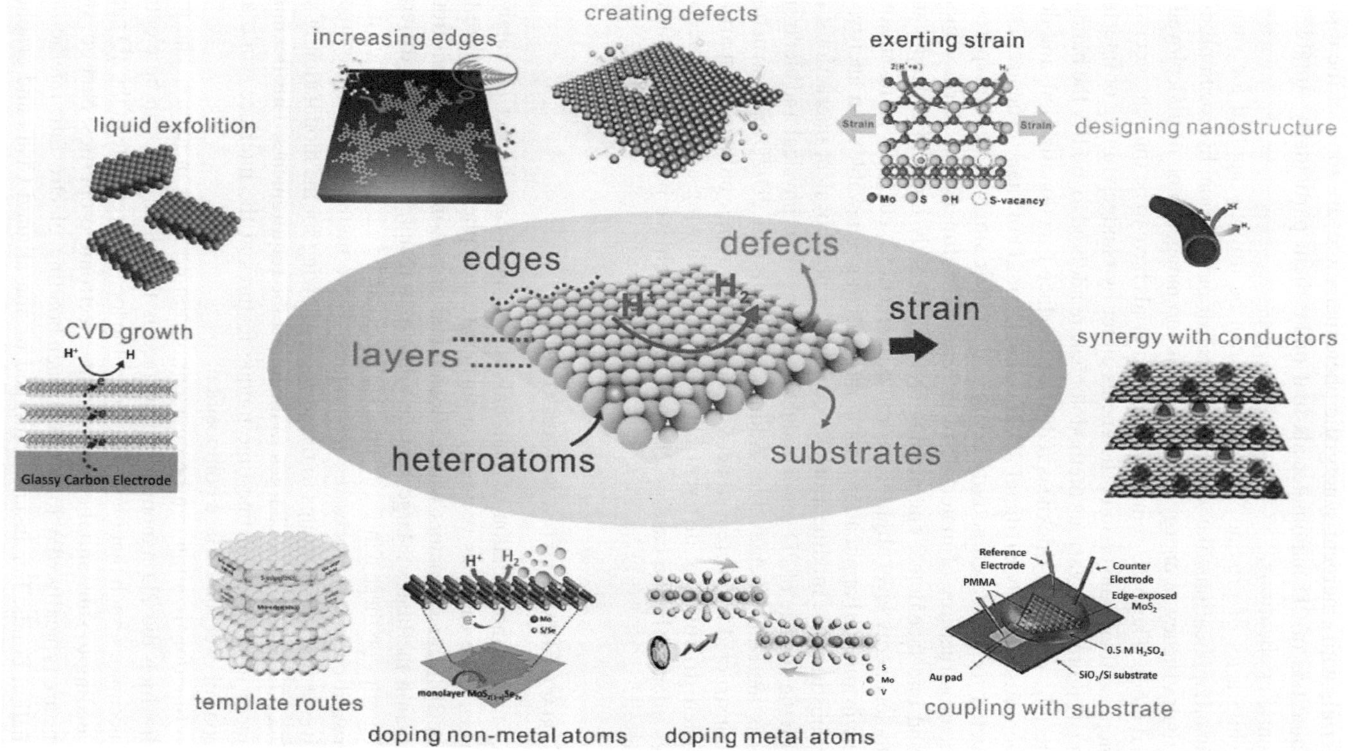

FIGURE 1.2　Strategies to improve catalytic performance in 2D nanomaterials (Chen et al. 2017). Copyright permission from the Royal Society of Chemistry.

cells as comparable to Pt catalysts (Wang, Yu, and Dai 2011; Wang et al. 2011). Similar reports are rGO modified with poly(3,4-ethylenedioxythiophene) polystyrene sulfonate (PEDOT: PSS) for enhanced optoelectronic applications (Sharma et al. 2021), organic molecule tetracyanoethylene adsorbed graphene sheets for ORR and oxygen evolution reactions (OER) (Zhao, Zhang, and Xia 2016), etc. At the same time, covalent interactions are typical bond formations with the surface atoms of the 2D semiconductors. Spontaneous grafting of nitrophenyl groups onto the carbon atoms of graphene, for example, can modify the local structure of graphene and, in fact, enhance its electrocatalytic H_2O_2 biosensing properties (Wang et al. 2013). MoS_2 covalently functionalized with a 1,2-dithiolane derivative followed by immobilization of Pd nanoparticle and preserving the integrity of MoS_2 basal planes showed superior ORR performance in the alkaline medium than its graphene counterpart or benchmark Pd/C catalyst (Perivoliotis et al. 2020). The introduction of such functionalization could enhance the solubility and avoid restacking nanosheets in common solvents and also allows the strategic designing of new hybrids (Figure 1.3).

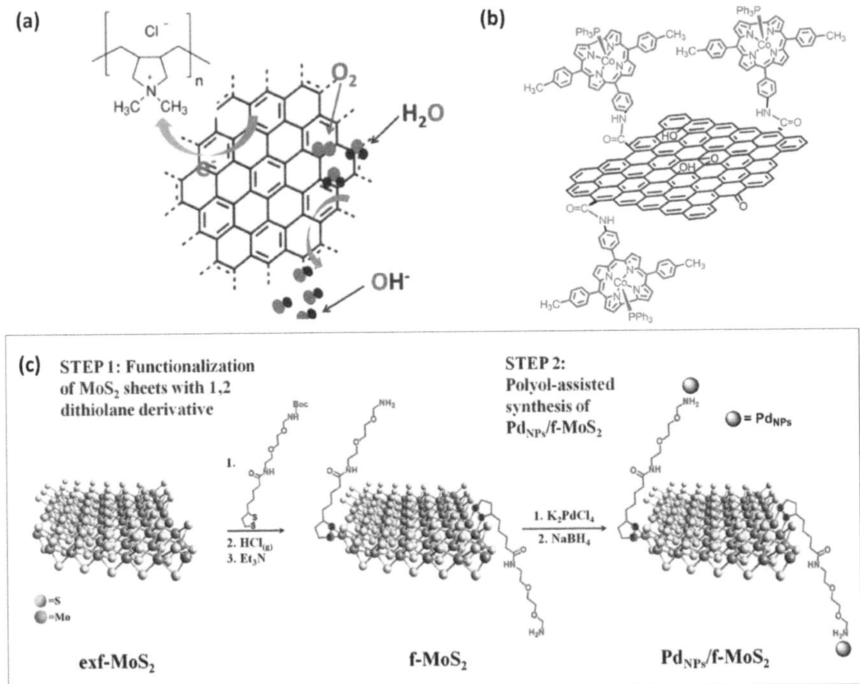

FIGURE 1.3   Illustrative scheme of (a) charge transfer and ORR in non-covalently linked PDDA absorbed graphene (Wang et al. 2011). Copyright permission from American Chemical Society. (b) Functionalized cobalt triarylcorrole covalently bonded with graphene oxide (Tang et al. 2017). Reprinted with permission from American Chemical Society. (c) Scheme for the synthesis of covalently functionalized MoS_2 with Pd immobilization (Perivoliotis et al. 2020). Copyright permission from Royal Society of Chemistry.

1.4.2　METAL AND NON-METAL DOPING

The approach of heteroatom doping is effective in adjusting the elemental composition and electrical features of 2D nanomaterials and to lower their adsorption barrier in the photocatalytic or electrocatalytic process for various environmental applications. Doping of surface heteroatoms may encourage the effect of electron-electron interactions and initiate electron transfer to improve electrical conductivity, even converting a portion of static surface sites into active sites. Doping defects are inserted into the catalyst's lattice due to the various radii of heteroatoms used, which impacts the catalytic properties of the 2D materials. The bandgaps of the 2D semiconductor will also get tuned by the extent of heteroatom doping. Dopants can either change the localized electronic states in the region of the prohibited bandgap or produce shallow electronic states close to the valence band maximum or conduction band minimum. For instance, N-doping in graphene causes the electrical band structure to shift upward (Kumar, Boukherroub, and Shankar 2018). N-doping in graphene results in the formation of pyridinic, pyrrolic, and quaternary nitrogen configurations, of which pyridinic nitrogen is the most beneficial in catalytic property enhancements due to the presence of lone pairs in the orbital. In addition, pyridinic nitrogen elevates the Fermi level over the Dirac point and raises the electronic DOSs, along with N atoms causing lattice distortions, all resulting in bandgap opening in N-doped graphene.

A practical and controllable technique for improving the catalytic activity of 2D nanocatalysts is heteroatom doping. The nature of doping can be substitutional, intercalation, electrostatic, or charge transfer, which determines the engineering of their properties. The n-type and p-type doping are usually substitutional. For the case of MoS_2, the metal atoms at the outer chalcogenide layer principally aid in selecting the metal atoms that have been integrated into the substitutional locations. In 2D material frameworks like graphene, non-metal-atom doping often involves single-element doping (N, B, O, S, F, etc.), dual-doping (BN-, SN-, etc.), and triple-doping (BNP, etc.). Metal-atom doping is mainly used to create single-atom catalysts on the 2D surfaces of 2D materials with a single Pt atom. B, N-co-doped rGO, which can be considered as a carbon-based peroxidase mimic, is demonstrated to show 1000-fold higher catalytic activity than that of undoped rGO (Figure 1.4) (Kim et al. 2019). Gaseous molecules, metals, and their compounds are the sources of charge transfer doping. In charge transfer doping, the host and dopant materials interact physically or chemically to form the contact, and the difference at their Fermi level determines the dopant activity concerning charge transfer direction. In comparison, the source of intercalation doping is usually H or alkali metal intercalations. The presence of vdW interlayers of 2D nanomaterials often facilitates such intercalations of foreign atoms.

Heteroatom doping in borophene can enhance the stability of the monolayer without restacking to a greater extent. Doping indeed affects the vacancy of the nearby boron atoms and changes from 4 to 6-coordinated atoms, and that vacancy decreases π and increases σ occupancies of neighboring B atoms, contributing toward its stability (Erakulan and Thapa 2022). Additionally, it is theoretically demonstrated that $\beta12$-borophene is higher OER performance by dual carbon doping. Density Functional Theory (DFT) calculations have shown the increasing free energy change for metal doping in borophene of the order: Co- = Fe- < Pd- < Pt- < Ni- < Rh- < Ru- <

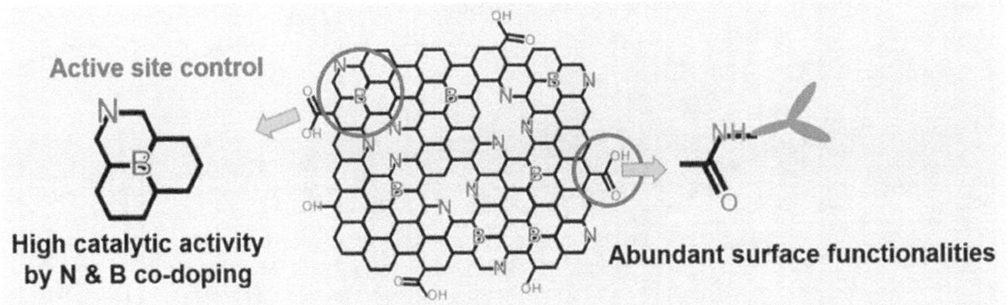

FIGURE 1.4 Schematic illustration of N,B-co-doped rGO (Kim et al. 2019). Copyright permission from American Chemical Society.

Ir- < Os-doped α-borophene, for electrocatalytic CO_2 conversion to useful chemical feedstocks (Xu et al. 2020). Similarly, it is predicted that Mo, Mn, Tc, and Cr doping on β12-borophene can exhibit superior performance in ammonia synthesis with low limiting potentials of −0.26, −0.32, −0.38, and −0.48 V, respectively (Xu, Yang, and Ganz 2021). In the case of silicene as well, the crystal lattice is found to have a very sensitive nature to the carrier concentration of dopants, however, the 2D material is evidenced to have a wide doping range (Cheng, Zhu, and Schwingenschlögl 2011).

1.4.3 Defect Engineering

Surface defects can significantly alter the local atomic structure, electronic structure, optical property, or electrical conductivity of a 2D semiconductor, which in turn affects the physicochemical characteristics and photocatalytic/electrocatalytic performance. Various surface defects, such as cation vacancies, anion vacancies, multi-vacancies, distortions, pits, and disorders, are studied to understand how the catalytic performance of the material is influenced in terms of their microstructures, atomic coordination, electronic states and electrical conductivity, carrier concentration, etc. Defects can be effectively introduced to a 2D semiconductor surface by chemical reduction, UV irradiation, vacuum activation/deactivation, rapid heating causing phase transitions, plasma etching, and lithium-induced treatments, etc., which can be precisely identified and quantified using characterization methods like EPR, Raman, TEM XPS, XAFS, etc. S-point defect-rich MoS_2 nanosheets prepared by ball milling with ascorbic acid are demonstrated with enhanced Cr(VI) removal during wastewater treatment (Luo et al. 2021). Ascorbic acid facilitates the S-stripping defects and increases the specific surface area of the 2D semiconductor. At the same time, the favored Mo-S dangling bonds trap the photogenerated holes paving charge-separation of electron-hole pairs during the photocatalytic processes. The ab-initio calculations also predict that S vacancies in MoS_2 nanosheets have extraordinary potency toward airborne elemental mercury adsorption (Zhao, Mu, et al. 2019). It has been demonstrated that local differences in stoichiometry, as well as metallic-like and structural defects, are connected with the variation in the electrical characteristics of MoS_2 throughout the surface of the same crystal (Addou, Colombo, and Wallace 2015) (Figure 1.5).

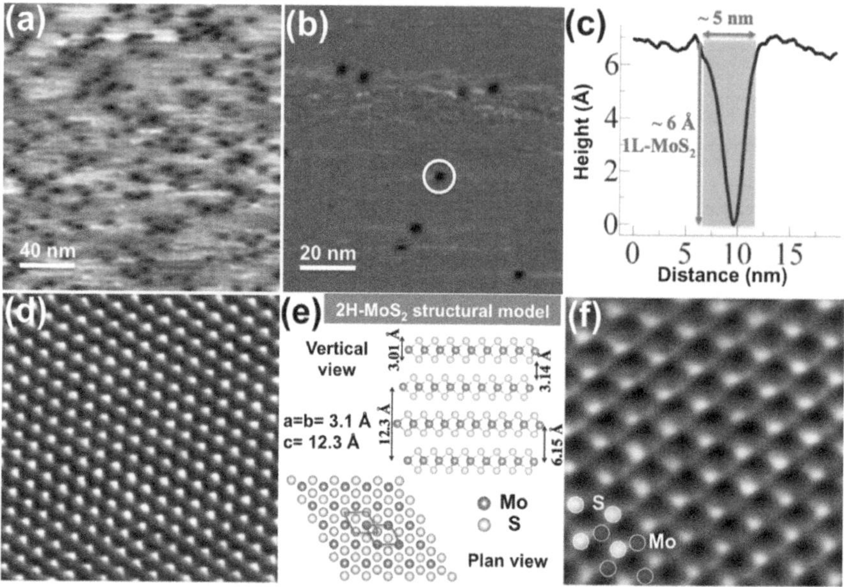

FIGURE 1.5 Topography of the MoS₂ crystal: (a) A large Scanning Tunnelling Microscopy (STM) image shows dark defects in high intensity. (b) STM picture captured in a region with low defect density. (c) Line profile over the dark defect shown in panel b. (d) High-resolution STM image in the defect-free section. (e) A schematic illustration of the bulk structure of 2H-MoS₂. (f) A well-resolved STM image displaying Mo- and S-layer structures (Addou, Colombo, and Wallace 2015). Copyright permission from American Chemical Society.

Graphene with single vacancy defects and ring defects has gained a lot of interest. According to certain simulation studies, the vacancy defects would lower Young's modulus of graphene. However, it is demonstrated that single vacancies at low concentrations might harden graphene, and that graphene will be weakened by high concentrations of single vacancies (Bhatt, Kim, and Kim 2022). Additionally, due to the dangling bond created by a single vacancy, single vacancies cause the local spin magnetic moment. Moreover, these atomic defects induce magnetism in graphene. Due to its symmetry and zero bandgap qualities, graphene's contact with a substrate can result in the charge or spin imbalances between the sublattices. This modifies the material's electrical and magnetic properties due to its changed lattice. The intrinsic features of graphene can also be tailored by structurally altering the edges using techniques like bond reconstruction, edge functionalization, passivation, doping, and strain. However, researchers are introducing single defects within perfect graphene (such as a tiny pore, a single step in a zigzag edge, or swapping out a single C atom for a Pt atom), which may have numerous uses, such as catalysis and electronic devices.

1.4.4 vDW Heterostructures

A precisely chosen LBL construction method can also be used to recombine individual atomic planes into designer heterostructures. The first of these heterostructures, known as "van der Waals", which is already extraordinarily complex, has lately been created and studied, exhibiting peculiar features and novel phenomena. With the development of 2D materials, novel heterostructures such as vdW stacks, which literary means laterally stitched 2D monolayers of different compounds, and more intricate layered and 3D architectures have never had such excellent prospects. Each layer in various 2D layered nanomaterials is covalently bonded without any dangling bonds, and vdW interactions weakly bind it to its neighboring layers. This removes the limitations of lattice matching and processing compatibility, making it possible to separate, mix, and match very dissimilar atomic layers to construct a variety of vdW heterostructures. New designs of electronic devices emerge by utilizing the novel properties in these vdW with the diverse layering of metals, semiconductors, or insulators.

The capacity to create atomically controlled heterostructures makes it possible to study the inherent properties of materials and uncharted physical phenomena at the most extreme 2D scale. For instance, graphene and MoS_2 resting on the h-BN dielectric layer have been found to have high mobility nearing theoretical values (Dean et al. 2010; Lee et al. 2013). The graphene/h-BN heterostructures have been used to study exotic physical phenomena such as the quantum Hall effect, Coulomb drag, and resonant tunneling.

1.5 CONCLUSIONS AND PERSPECTIVES

2D nanosheets can now be molecularly tuned for the creation of hybrid materials, thanks to advancements in their synthesis. It will be feasible to functionalize these nanosheets with molecules after their optical, thermal, electrical, and mechanical properties have been thoroughly explored. This will usher in a new era of materials with unrivaled uses. These 2D nanosheets might construct heterostructures by fusing with various biomolecules and piling them on top of one another for vdW heterostructures. Due to their exceptional qualities, 2D nanosheets are now a rising star in the nanotechnology industry. As it may be easily functionalized via covalent and non-covalent interactions, the research undertaken so far has shown encouraging findings for its usage in environmental applications. Additionally, the huge specific surface area of 2D nanomaterials promotes the deployment of those materials in surface-active applications like sensing and catalysis. Also, the high anisotropy and atomic thickness of 2D materials provide them excellent mechanical flexibility and optical transparency, opening up numerous potentials for developing wearable and optoelectronic devices based on 2D materials. Furthermore, without being constrained by lattice matching or processing compliance, dissimilar 2D materials can be combined to create hybrid materials, providing synergistic effects that are advantageous for a variety of applications.

There are still difficulties in the realm of 2D materials despite the great advancements. For ultimate industrialization, large-scale manufacture of 2D materials of high quality and precise structure has not yet been achieved. It is crucial to achieve perfect control over their composition, vacancy, strain, doping, defects, crystal phase, lateral dimensions, and surface properties to further reveal the relationship between their structural characteristics and properties. In addition, various kinds of 2D nanomaterials, like 2D metal nanomaterials and 2D perovskites, are thought to exhibit attractive qualities and need further study due to the considerable advancement in layered 2D materials. Given our current accomplishments and the tireless efforts of numerous research domains, it does seem inevitable that we need to eventually attain significant milestones of commercializing 2D nanomaterials in our everyday lives.

REFERENCES

Addou, Rafik, Luigi Colombo, and Robert M. Wallace. 2015. Surface defects on natural MoS_2. *ACS Applied Materials & Interfaces* 7 (22):11921–11929.

Allgayer, Raphaela, Nariman Yousefi, and Nathalie Tufenkji. 2020. Graphene oxide sponge as adsorbent for organic contaminants: comparison with granular activated carbon and influence of water chemistry. *Environmental Science: Nano* 7 (9):2669–2680.

Araújo, Mariana P., O. S. G. P. Soares, A. J. S. Fernandes, M. F. R. Pereira, and C. Freire. 2017. Tuning the surface chemistry of graphene flakes: new strategies for selective oxidation. *RSC Advances* 7 (23):14290–14301.

Babenko, Vitaliy, George Lane, Antal A. Koos, et al. 2017. Time dependent decomposition of ammonia borane for the controlled production of 2D hexagonal boron nitride. *Scientific Reports* 7 (1):14297.

Bhatt, Mahesh Datt, Heeju Kim, and Gunn Kim. 2022. Various defects in graphene: a review. *RSC Advances* 12 (33):21520–21547.

Bianco, Elisabeth, Sheneve Butler, Shishi Jiang, Oscar D. Restrepo, Wolfgang Windl, and Joshua E. Goldberger. 2013. Stability and exfoliation of germanane: a germanium graphane analogue. *ACS Nano* 7 (5):4414–4421.

Brodie, Benjamin Collins. 1859. XIII. On the atomic weight of graphite. *Philosophical Transactions of the Royal Society of London* 149:249–259.

Cassabois, G., P. Valvin, and B. Gil. 2016. Hexagonal boron nitride is an indirect bandgap semiconductor. *Nature Photonics* 10 (4):262–266.

Castellanos-Gomez, Andres, Leonardo Vicarelli, Elsa Prada, et al. 2014. Isolation and characterization of few-layer black phosphorus. *2D Materials* 1 (2):025001.

Chen, Ximin, Yingqing Zhan, Ao Sun, et al. 2022. Anchoring the TiO_2@crumpled graphene oxide core–shell sphere onto electrospun polymer fibrous membrane for the fast separation of multi-component pollutant-oil–water emulsion. *Separation and Purification Technology* 298:121605.

Chen, Yunxu, Kena Yang, Bei Jiang, Jiaxu Li, Mengqi Zeng, and Lei Fu. 2017. Emerging two-dimensional nanomaterials for electrochemical hydrogen evolution. *Journal of Materials Chemistry* A 5 (18):8187–8208.

Cheng, Y. C., Z. Y. Zhu, and U. Schwingenschlögl. 2011. Doped silicene: evidence of a wide stability range. *Europhysics Letters* 95 (1):17005.

Choi, Byung-jin. 1999. Chemical vapor deposition of hexagonal boron nitride films in the reduced pressure. *Materials Research Bulletin* 34 (14):2215–2220.

Choi, Wonbong, Nitin Choudhary, Gang Hee Han, Juhong Park, Deji Akinwande, and Young Hee Lee. 2017. Recent development of two-dimensional transition metal dichalcogenides and their applications. *Materials Today* 20 (3):116–130.

Cunningham, Graeme, Mustafa Lotya, Clotilde S. Cucinotta, et al. 2012. Solvent exfoliation of transition metal dichalcogenides: dispersibility of exfoliated nanosheets varies only weakly between compounds. *ACS Nano* 6 (4):3468–3480.

Dávila, M. E., L. Xian, S. Cahangirov, A. Rubio, and G. Le Lay. 2014. Germanene: a novel two-dimensional germanium allotrope akin to graphene and silicene. *New Journal of Physics* 16 (9):095002.

Dean, C. R., A. F. Young, I. Meric, et al. 2010. Boron nitride substrates for high-quality graphene electronics. *Nature Nanotechnology* 5 (10):722–726.

Dreyer, Daniel R., Sungjin Park, Christopher W. Bielawski, and Rodney S. Ruoff. 2010. The chemistry of graphene oxide. *Chemical Society Reviews* 39 (1):228–240.

Erakulan, E.S., and Ranjit Thapa. 2022. Origin of pure and C doped borophene stability and its activity for OER. *Applied Surface Science* 574:151613.

Eswaraiah, Varrla, Sasidharannair Sasikaladevi Jyothirmayee Aravind, and Sundara Ramaprabhu. 2011. Top down method for synthesis of highly conducting graphene by exfoliation of graphite oxide using focused solar radiation. *Journal of Materials Chemistry* 21 (19):6800–6803.

Fernández-Merino, M. J., L. Guardia, J. I. Paredes, et al. 2010. Vitamin C is an ideal substitute for hydrazine in the reduction of graphene oxide suspensions. *The Journal of Physical Chemistry C* 114 (14):6426–6432.

Foller, Tobias, Huanting Wang, and Rakesh Joshi. 2022. Rise of 2D materials-based membranes for desalination. *Desalination* 536:115851.

Fu, Xianliang, Yingfei Hu, Yunguang Yang, We Liu, and Shifu Chen. 2013. Ball milled h-BN: an efficient holes transfer promoter to enhance the photocatalytic performance of TiO_2. *Journal of Hazardous Materials* 244–245:102–110.

Gautam, Chandkiram, and Selvam Chelliah. 2021. Methods of hexagonal boron nitride exfoliation and its functionalization: covalent and non-covalent approaches. *RSC Advances* 11 (50):31284–31327.

Geim, A. K., and I. V. Grigorieva. 2013. Van der Waals heterostructures. *Nature* 499 (7459):419–425.

Gogotsi, Yury, and Babak Anasori. 2019. The rise of MXenes. *ACS Nano* 13 (8):8491–8494.

Gupta, Amit, Vaishali Arunachalam, and Sukumaran Vasudevan. 2015. Water dispersible, positively and negatively charged MoS_2 nanosheets: surface chemistry and the role of surfactant binding. *The Journal of Physical Chemistry Letters* 6 (4):739–744.

Hidalgo, A., V. Makarov, G. Morell, and B. R. Weiner. 2013. High-yield synthesis of cubic and hexagonal boron nitride nanoparticles by laser chemical vapor decomposition of borazine. *Dataset Papers in Nanotechnology* 2013:281672.

Hope, Michael A., Alexander C. Forse, Kent J. Griffith, et al. 2016. NMR reveals the surface functionalisation of Ti_3C_2 MXene. *Physical Chemistry Chemical Physics* 18 (7):5099–5102.

Hu, Jun, and Zuoguo Yang. 2021. Layer-by-layer self-assembly preparation and desalination performance of graphene oxide membrane. *Water Supply* 22 (1):126–136.

Hummers, William S., Jr., and Richard E. Offeman. 1958. Preparation of graphitic oxide. *Journal of the American Chemical Society* 80 (6):1339–1339.

Iijima, Sumio. 1991. Helical microtubules of graphitic carbon. *Nature* 354 (6348):56–58.

Jang, Sung Kyu, Jiyoun Youn, Young Jae Song, and Sungjoo Lee. 2016. Synthesis and characterization of hexagonal boron nitride as a gate dielectric. *Scientific Reports* 6 (1):30449.

Joshi, R. K., P. Carbone, F. C. Wang, et al. 2014. Precise and ultrafast molecular sieving through graphene oxide membranes. *Science* 343 (6172):752–754.

Kavitha, M. K., Pramod Gopinath, and Honey John. 2015. Reduced graphene oxide–ZnO self-assembled films: tailoring the visible light photoconductivity by the intrinsic defect states in ZnO. *Physical Chemistry Chemical Physics* 17 (22):14647–14655.

Kim, Min Su, Seongyeon Cho, Se Hun Joo, et al. 2019. N- and B-codoped graphene: a strong candidate to replace natural peroxidase in sensitive and selective bioassays. *ACS Nano* 13 (4):4312–4321.

Koepke, Justin C., Joshua D. Wood, Yaofeng Chen, et al. 2016. Role of pressure in the growth of hexagonal boron nitride thin films from ammonia-borane. *Chemistry of Materials* 28 (12):4169–4179.

Kroto, H. W., J. R. Heath, S. C. O'Brien, R. F. Curl, and R. E. Smalley. 1985. C60: buckminsterfullerene. *Nature* 318 (6042):162–163.

Kumar, Pawan, Rabah Boukherroub, and Karthik Shankar. 2018. Sunlight-driven water-splitting using two-dimensional carbon based semiconductors. *Journal of Materials Chemistry A* 6 (27):12876–12931.

Lee, Gwan-Hyoung, Young-Jun Yu, Xu Cui, et al. 2013. Flexible and transparent MoS_2 field-effect transistors on hexagonal boron nitride-graphene heterostructures. *ACS Nano* 7 (9):7931–7936.

Li, Qun, Tao Yang, Qifan Yang, Fei Wang, Kuo-Chih Chou, and Xinmei Hou. 2016. Porous hexagonal boron nitride whiskers fabricated at low temperature for effective removal of organic pollutants from water. *Ceramics International* 42 (7):8754–8762.

Li, Shaocheng, Xianlang Lu, Yanda Lou, Kejun Liu, and Benxue Zou. 2021. The synthesis and characterization of h-BN nanosheets with high yield and crystallinity. *ACS Omega* 6 (42):27814–27822.

Li, Zizhen, Xiangchao Meng, and Zisheng Zhang. 2018. Recent development on MoS_2-based photocatalysis: a review. *Journal of Photochemistry and Photobiology C: Photochemistry Reviews* 35:39–55.

Lin, Yi, Tiffany V. Williams, and John W. Connell. 2010. Soluble, exfoliated hexagonal boron nitride nanosheets. *The Journal of Physical Chemistry Letters* 1 (1):277–283.

Liu, Pengchao, Junjun Hou, Yi Zhang, Lianshan Li, Xiaoquan Lu, and Zhiyong Tang. 2020. Two-dimensional material membranes for critical separations. *Inorganic Chemistry Frontiers* 7 (13):2560–2581.

Liu, Qiuwen, Cheng Chen, Man Du, et al. 2018. Porous hexagonal boron nitride sheets: effect of hydroxyl and secondary amino groups on photocatalytic hydrogen evolution. *ACS Applied Nano Materials* 1 (9):4566–4575.

Lopez-Bezanilla, Alejandro, Jingsong Huang, Humberto Terrones, and Bobby G. Sumpter. 2011. Boron nitride nanoribbons become metallic. *Nano Letters* 11 (8):3267–3273.

Lowe, Sean E., Ge Shi, Yubai Zhang, et al. 2019. The role of electrolyte acid concentration in the electrochemical exfoliation of graphite: mechanism and synthesis of electrochemical graphene oxide. *Nano Materials Science* 1 (3):215–223.

Lozovoy, Kirill A., Ihor I. Izhnin, Andrey P. Kokhanenko, et al. 2022. Single-element 2D materials beyond graphene: methods of epitaxial synthesis. *Nanomaterials* 12 (13):2221.

Luo, Ni, Cheng Chen, Dingming Yang, Wenyuan Hu, and Faqin Dong. 2021. S defect-rich ultrathin 2D MoS_2: the role of S point-defects and S stripping-defects in the removal of Cr(VI) via synergistic adsorption and photocatalysis. *Applied Catalysis B: Environmental* 299:120664.

Maleki, Mahdi, Ali Beitollahi, and Mohammadreza Shokouhimehr. 2015. Template-free synthesis of porous boron nitride using a single source precursor. *RSC Advances* 5 (58):46823–46828.

Mannix, Andrew J., Xiang-Feng Zhou, Brian Kiraly, et al. 2015. Synthesis of borophenes: anisotropic, two-dimensional boron polymorphs. *Science* 350 (6267):1513–1516.

Masud, Arvid, Chi Zhou, and Nirupam Aich. 2021. Emerging investigator series: 3D printed graphene-biopolymer aerogels for water contaminant removal: a proof of concept. *Environmental Science: Nano* 8 (2):399–414.

Mattevi, Cecilia, Goki Eda, Stefano Agnoli, et al. 2009. Evolution of electrical, chemical, and structural properties of transparent and conducting chemically derived graphene thin films. *Advanced Functional Materials* 19 (16):2577–2583.

Naguib, Michael, Murat Kurtoglu, Volker Presser, et al. 2011. Two-dimensional nanocrystals produced by exfoliation of Ti_3AlC_2. *Advanced Materials* 23 (37):4248–4253.

Novoselov, K. S., A. K. Geim, S. V. Morozov, et al. 2004. Electric field effect in atomically thin carbon films. 306 (5696):666–669.

Pei, Songfeng, Qinwei Wei, Kun Huang, Hui-Ming Cheng, and Wencai Ren. 2018. Green synthesis of graphene oxide by seconds timescale water electrolytic oxidation. *Nature Communications* 9 (1):145.

Pei, Songfeng, Jinping Zhao, Jinhong Du, Wencai Ren, and Hui-Ming Cheng. 2010. Direct reduction of graphene oxide films into highly conductive and flexible graphene films by hydrohalic acids. *Carbon* 48 (15):4466–4474.

Peng, Qiuming, Jianxin Guo, Qingrui Zhang, et al. 2014. Unique lead adsorption behavior of activated hydroxyl group in two-dimensional titanium carbide. *Journal of the American Chemical Society* 136 (11):4113–4116.

Perivoliotis, Dimitrios K., Yuta Sato, Kazu Suenaga, and Nikos Tagmatarchis. 2020. Covalently functionalized layered MoS_2 supported Pd nanoparticles as highly active oxygen reduction electrocatalysts. *Nanoscale* 12 (35):18278–18288.

Rai, V., M. K. Kavitha, and M. Jaiswal. 2015. Correlating chemical structure and charge transport in reduced graphene oxide for transparent conductor and interconnect applications. *IEEE International Symposium on Nanoelectronic and Information Systems*, 21–23 Dec. 2015.

Rasch, Florian, Fabian Schütt, Lena M. Saure, et al. 2019. Wet-chemical assembly of 2D nanomaterials into lightweight, microtube-shaped, and macroscopic 3D networks. *ACS Applied Materials & Interfaces* 11 (47):44652–44663.

Saba, N., and M. Jawaid. 2018. 4 - Energy and environmental applications of graphene and its derivatives. In *Polymer-based Nanocomposites for Energy and Environmental Applications*, edited by M. Jawaid, and M. M. Khan: Woodhead Publishing.

Seenivasan, Rajesh, Woo-Jin Chang, and Sundaram Gunasekaran. 2015. Highly sensitive detection and removal of lead ions in water using cysteine-functionalized graphene oxide/polypyrrole nanocomposite film electrode. *ACS Applied Materials & Interfaces* 7 (29):15935–15943.

Sharma, Sandeep, Karamvir Singh, Sandeep Kumar, et al. 2021. Fabrication of reduced graphene oxide modified poly(3,4-ethylenedioxythiophene) polystyrene sulfonate based transparent conducting electrodes for flexible optoelectronic application. *SN Applied Sciences* 3 (1):26.

Shin, Hyeon-Jin, Ki Kang Kim, Anass Benayad, et al. 2009. Efficient reduction of graphite oxide by sodium borohydride and its effect on electrical conductance. *Advanced Functional Materials* 19 (12):1987–1992.

Song, Jun-ho, Hye-Weon Yu, Moon-Ho Ham, and In S. Kim. 2018. Tunable ion sieving of graphene membranes through the control of nitrogen-bonding configuration. *Nano Letters* 18 (9):5506–5513.

Swain, Akshaya Kumar, and Dhirendra Bahadur. 2014. Enhanced stability of reduced graphene oxide colloid using cross-linking polymers. *The Journal of Physical Chemistry C* 118 (18):9450–9457.

Tan, Chaoliang, Xiaoying Qi, Zhengdong Liu, et al. 2015. Self-assembled chiral nanofibers from ultrathin low-dimensional nanomaterials. *Journal of the American Chemical Society* 137 (4):1565–1571.

Tang, Jijun, Zhongping Ou, Rui Guo, et al. 2017. Functionalized cobalt triarylcorrole covalently bonded with graphene oxide: a selective catalyst for the two- or four-electron reduction of oxygen. *Inorganic Chemistry* 56 (15):8954–8963.

Velusamy, Dhinesh, Richard Kim, Soonyoung Cha, et al. 2015. Flexible transition metal dichalcogenide nanosheets for band-selective photodetection. *Nature Communications* 6:8063.

Viculis, Lisa M., Julia J. Mack, and Richard B. Kaner. 2003. A chemical route to carbon nano-scrolls. *Science* 299 (5611):1361–1361.

Vogt, Patrick, Paola De Padova, Claudio Quaresima, et al. 2012. Silicene: compelling experimental evidence for graphenelike two-dimensional silicon. *Physical Review Letters* 108 (15):155501.

Voiry, Damien, Jieun Yang, Jacob Kupferberg, et al. 2016. High-quality graphene via microwave reduction of solution-exfoliated graphene oxide. *Science* 353 (6306):1413–1416.

Wang, Jiemin, Liangzhu Zhang, Lifeng Wang, Weiwei Lei, and Zhong-Shuai Wu. 2022. Two-dimensional boron nitride for electronics and energy applications. *Energy & Environmental Materials* 5 (1):10–44.

Wang, Li, Yinjian Ye, Xingping Lu, et al. 2013. Prussian blue nanocubes on nitrobenzene-functionalized reduced graphene oxide and its application for H_2O_2 biosensing. *Electrochimica Acta* 114:223–232.

Wang, Ning, Guang Yang, Haixu Wang, Rong Sun, and Ching-Ping Wong. 2018. Visible light-responsive photocatalytic activity of boron nitride incorporated composites. *Frontiers in Chemistry* 6:1–12.

Wang, Shuangyin, Dingshan Yu, and Liming Dai. 2011. Polyelectrolyte functionalized carbon nanotubes as efficient metal-free electrocatalysts for oxygen reduction. *Journal of the American Chemical Society* 133 (14):5182–5185.

Wang, Shuangyin, Dingshan Yu, Liming Dai, Dong Wook Chang, and Jong-Beom Baek. 2011. Polyelectrolyte-functionalized graphene as metal-free electrocatalysts for oxygen reduction. *ACS Nano* 5 (8):6202–6209.

Wang, Zhongying, and Baoxia Mi. 2017. Environmental applications of 2D molybdenum disulfide (MoS_2) nanosheets. *Environmental Science & Technology* 51 (15):8229–8244.

Xie, Junfeng, Jiajia Zhang, Shuang Li, et al. 2013. Controllable disorder engineering in oxygen-incorporated MoS_2 ultrathin nanosheets for efficient hydrogen evolution. *Journal of the American Chemical Society* 135 (47):17881–17888.

Xu, Lu, Li-Ming Yang, and Eric Ganz. 2021. Electrocatalytic reduction of N_2 using metal-doped borophene. *ACS Applied Materials & Interfaces* 13 (12):14091–14101.

Xu, Xuejian, Xiuli Hou, Jiajie Lu, Peng Zhang, Beibei Xiao, and Jianli Mi. 2020. Metal-doped two-dimensional borophene nanosheets for the carbon dioxide electrochemical reduction reaction. *The Journal of Physical Chemistry C* 124 (44):24156–24163.

Xue, Lili, Binzeng Lu, Zhong-Shuai Wu, et al. 2014. Synthesis of mesoporous hexagonal boron nitride fibers with high surface area for efficient removal of organic pollutants. *Chemical Engineering Journal (Lausanne)* 243:494–499.

Yan, Mengyu, Fengchao Wang, Chunhua Han, et al. 2013. Nanowire templated semihollow bicontinuous graphene scrolls: designed construction, mechanism, and enhanced energy storage performance. *Journal of the American Chemical Society* 135 (48):18176–18182.

Yar, Asfand, John Ojur Dennis, Amina Yasin, et al. 2021. The solar reduction of graphene oxide on a large scale for high density electrochemical energy storage. *Sustainable Energy & Fuels* 5 (10):2724–2733.

Ye, Huijian, Tiemei Lu, Chunfeng Xu, Bo Han, Nan Meng, and Lixin Xu. 2018. Liquid-phase exfoliation of hexagonal boron nitride into boron nitride nanosheets in common organic solvents with hyperbranched polyethylene as stabilizer. *Macromolecular Chemistry and Physics* 219 (6):1700482.

Ying, Yulong, Yu Liu, Xinyu Wang, et al. 2015. Two-dimensional titanium carbide for efficiently reductive removal of highly toxic chromium(VI) from water. *ACS Applied Materials & Interfaces* 7 (3):1795–1803.

Zhang, Yi, Luyao Zhang, and Chongwu Zhou. 2013. Review of chemical vapor deposition of graphene and related applications. *Accounts of Chemical Research* 46 (10):2329–2339.

Zhao, Haitao, Xueliang Mu, Chenghang Zheng, et al. 2019. Structural defects in 2D MoS_2 nanosheets and their roles in the adsorption of airborne elemental mercury. *Journal of Hazardous Materials* 366:240–249.

Zhao, Jinping, Songfeng Pei, Wencai Ren, Libo Gao, and Hui-Ming Cheng. 2010. Efficient preparation of large-area graphene oxide sheets for transparent conductive films. *ACS Nano* 4 (9):5245–5252.

Zhao, Zeyuan, Shuainan Ni, Xiang Su, Yun Gao, and Xiaoqi Sun. 2019. Thermally reduced graphene oxide membrane with ultrahigh rejection of metal ions' separation from water. *ACS Sustainable Chemistry & Engineering* 7 (17):14874–14882.

Zhao, Zhenghang, Lipeng Zhang, and Zhenhai Xia. 2016. Electron transfer and catalytic mechanism of organic molecule-adsorbed graphene nanoribbons as efficient catalysts for oxygen reduction and evolution reactions. *The Journal of Physical Chemistry C* 120 (4):2166–2175.

Zhong, Mianzeng, Qinglin Xia, Longfei Pan, et al. 2018. Thickness-dependent carrier transport characteristics of a new 2D elemental semiconductor: black arsenic. *Advanced Functional Materials* 28 (43):1802581.

Zhou, Dong, Haibo Shu, Chenli Hu, Li Jiang, Pei Liang, and Xiaoshuang Chen. 2018. Unveiling the growth mechanism of MoS$_2$ with chemical vapor deposition: from two-dimensional planar nucleation to self-seeding nucleation. *Crystal Growth & Design* 18 (2):1012–1019.

Zhou, Fang, Jingang Yu, and Xinyu Jiang. 2017. 3D porous graphene synthesised using different hydrothermal treatment times for the removal of lead ions from an aqueous solution. *Micro & Nano Letters* 12 (5):308–311.

Zhou, Ming, Yuling Wang, Yueming Zhai, et al. 2009. Controlled synthesis of large-area and patterned electrochemically reduced graphene oxide films. *Chemistry – A European Journal* 15 (25):6116–6120.

Zhu, Feng-feng, Wei-jiong Chen, Yong Xu, et al. 2015. Epitaxial growth of two-dimensional stanene. *Nature Materials* 14 (10):1020–1025.

Zhuang, Pengyu, Zhongya Guo, Shuang Wang, et al. 2021. Interfacial hydrothermal assembly of three-dimensional lamellar reduced graphene oxide aerogel membranes for water self-purification. *ACS Omega* 6 (45):30656–30665.

2 Synthesis, Growth, and Chemical Functionalization of 2D Semiconductors and Heterostructures

Priyanka Ganguly, Devagi Kanakaraju and Lim Ying Chin

2.1 INTRODUCTION

Two-dimensional (2D) semiconductors have attracted great attention among researchers due to their properties which have found numerous applications in electronics, optical, and optoelectronics (Lee et al. 2016). Since the discovery of graphene in 2004 with unique properties (Velický and Toth 2017), 2D materials have led to a revolution in materials with ultrathin dimensions. A search on the Web of Science demonstrates an exponential increase in the number of publications since the year 2000 (Figure 2.1a). Applications for 2D semiconductors span a variety of industries, including electrical and electronic, materials science, energy fuels, and engineering (Figure 2.1b).

In the architectures of the various reported 2D nanomaterials, very intriguing physicochemical events related to electron transfers and bandgap modulations have been observed, clearly distinguishing them from their bulk counterparts (Pillai and Ganguly 2021). Examples of 2D materials are graphene, metal and non-metal chalcogenides, oxides, hydroxides, halides, black phosphorus (BP), covalent organic frameworks, silicates, perovskites, and boron nitride (BN) (Fatima et al. 2022). Contrary to the characteristics of their bulk parent materials, 2D semiconductors are endowed with distinctive and interesting properties due to atomic scale thickness (Song, Hu, and Zeng 2013). Intriguing characteristics such as unique electronic band structure, high carrier mobility, thermal conductivity, and high transparency made graphene an excellent 2D material. Although graphene stands out as a 2D material, it does not have a bandgap (or gapless band structure) which implies that this material is unfit to act as a semiconductor (Hu et al. 2019). Thus, the development of graphene-based 2D materials or other 2D-based materials has been a major focus. The current trend shows that 2D materials of various forms of conductivity are combined into van der

DOI: 10.1201/9781003343899-2

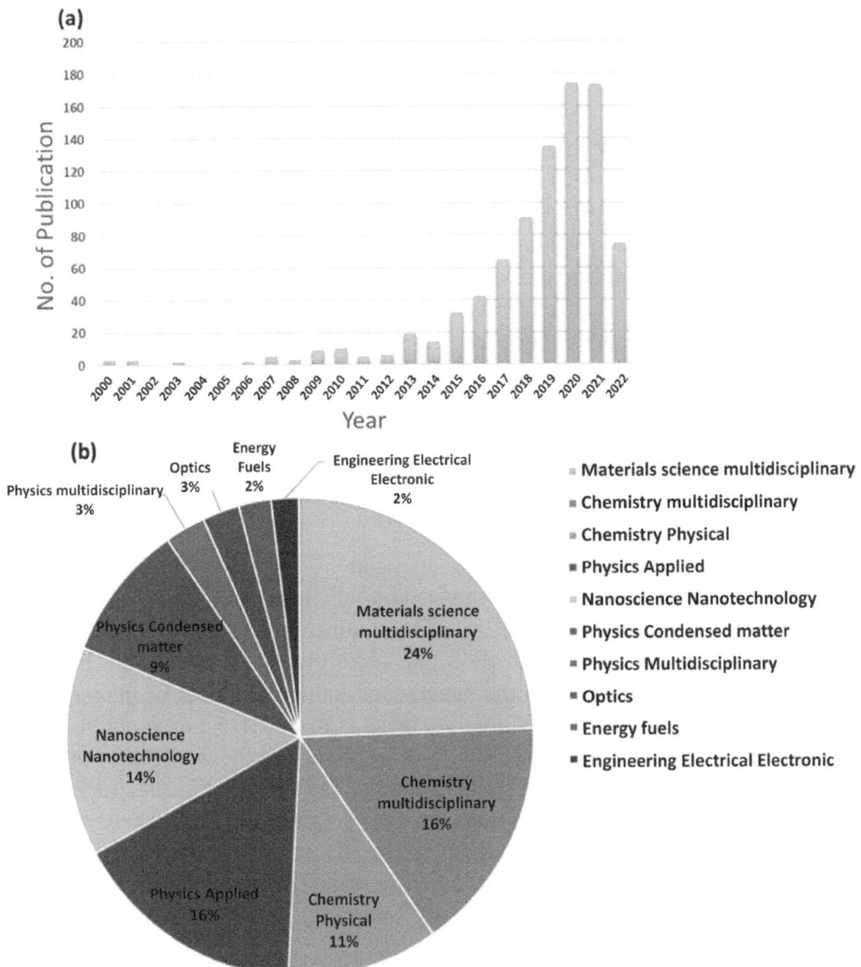

FIGURE 2.1 (a) Number of publications over years for the keyword search '2D Semiconductors' on the Web of Science, (b) fields of 2D semiconductors.

Waals heterostructures in creating electronic devices with specific properties (Song et al. 2018).

In this chapter, we discuss the different physical and chemical properties of some of the most known 2D nanomaterials. The further section of the chapter discusses wide variety of synthesis protocols adapted to synthesize 2D nanosheets of varying levels of size and thickness. The chapter emphasizes on some of the state-of-the-art methods adapted and further discusses their limitations as well. Moreover, the stakes for the upcoming research and challenges ahead in terms of energy and environmental sustainability are discussed and the impact of the research on 2D nanomaterials can play a fundamental role.

2.2 PROPERTIES OF 2D SEMICONDUCTORS

2D materials are crystalline solids having a large ratio between their lateral dimensions (between 1 and 10,000 m) and thickness (under 1 nm) (Velický and Toth 2017). 2D materials are a class of substances in which layers of a single atom are joined together by van der Waals interactions in the vertical direction, as opposed to the horizontal covalent bonds that typically join layers together (Velický and Toth 2017). The intriguing characteristics of monolayer or few-layer 2D materials include high surface area or surface-to-volume ratio, lack of dangling bonds, surface state, distinctive spin-orbital coupling properties, and quantum spin Hall effects. Also, these materials possess a wide range of electronic properties such as high carrier mobility, unique electronic and optical band structures, and excellent light-matter interaction (Zhang et al. 2019). 2D crystals are stabilized in-plane by strong covalent bonds, whereas the stack is held together by relatively weak van-der-Waals-like forces (Geim and Grigorieva 2013). Figure 2.2 displays some of the widely researched 2D nanomaterials. The applications of these materials extend across various broad applications notably, in catalysis, optoelectronics, energy storage, biomedicine, sensors, supercapacitors, batteries, *etc.*

In particular, photocatalysis and its applications such as hydrogen generation, oxygen reduction reaction, disinfection, and water remediation are well reported over the course of the past decade. When heterojunction and 2D structure are combined to create a 2D/2D heterojunction photocatalyst, their benefits can be assimilated and integrated to achieve the superposed enhanced effect of photocatalytic performance (Zhu et al. 2021). Numerous benefits of 2D/2D heterojunction include effective charge separation, plentiful light absorption, and the availability of a large number of active sites. One example of a popular 2D photocatalyst is graphitic carbon nitride (g-C_3N_4). Studies have shown that coupling other 2D materials with g-C_3N_4 can produce g-C_3N_4-based 2D/2D heterojunctions with excellent properties. Chen et al. (2021) fabricated a 2D/2D heterostructure of BP/bismuth tungstate (BP/ Bi_2WO_6) by modification of ultrasmall platinum (Pt) nanoparticles for photocatalytic CO_2 reduction. The CO and H_2 generation rates were 20.5 and 16.8 µmol $g^{-1}h^{-1}$, respectively, which are significantly greater than BP-based materials. The observed enhanced photocatalytic chemical oxygen demand reduction was due to the unique optical properties of BP nanosheets upon modification with Pt and enhanced charge transfer efficiency. Similarly, exploring their abilities for energy storage and exploring their morphological tuning and controlling the dimensions resulted in exhibiting interesting features and improvements.

In this section, we intend to discuss a few of the most used 2D nanomaterials. The properties of graphene, BP, hexagonal boron nitride (h-BN), transition metal dichalcogenides (TMDs), and oxides are discussed.

2.2.1 GRAPHENE

Graphene is one of the most famous 2D nanomaterials all of them and is used for various applications. The carbon atoms arranged in an sp^2 configuration in a honeycomb structure give them unique physical and chemical properties (Pillai and Ganguly

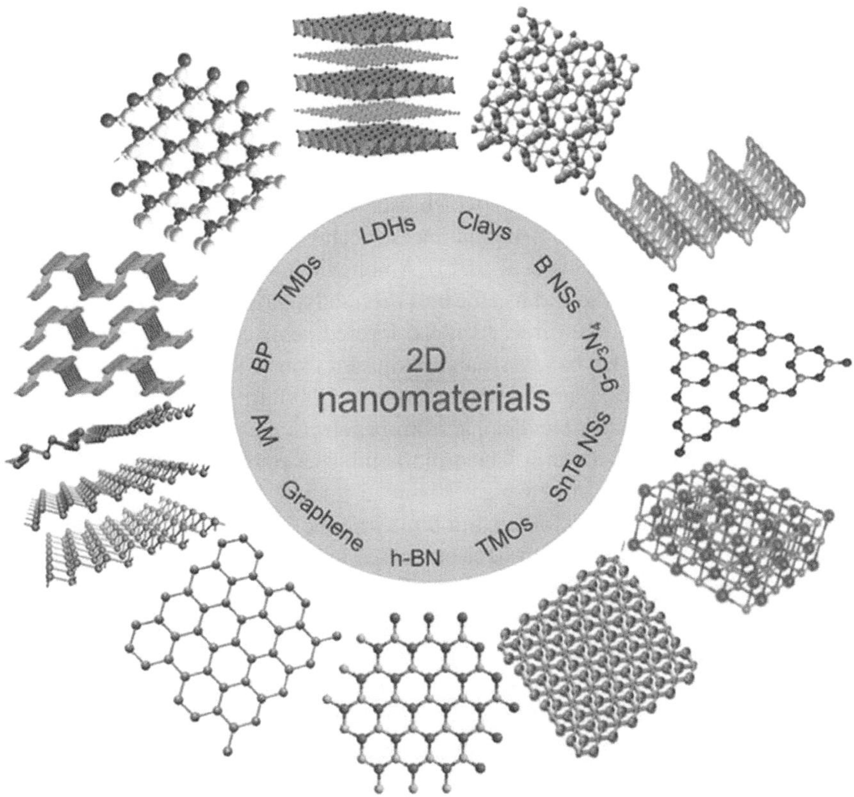

FIGURE 2.2 Some of the most well-known 2D nanomaterials. Reproduced with the permission of ref (Hu et al. 2019).

2021). Exfoliation and several other top-down synthesis approaches have been used to improve the synthesis process and enhance the yield of graphene (Nicolosi et al. 2013). Synthesizing single and even multiple layers of graphene nanomaterials of high quality is quite challenging. Common methods such as chemical vapor deposition (CVD) and other etching techniques as discussed in the following section are some of the methods used to produce high-quality graphene sheets (Wang et al. 2012). Due to its inherent high mobility and zero bandgap, graphene photodetector demonstrated exceptionally high operating frequency and very wide detection wavelength range. However, compared to most conventional semiconductor photodetectors, it has an extremely poor photo responsiveness due to its zero-bandgap.

The development of more adaptable and tunable 2D graphene-like materials or alternatives with a larger diversity of composition, structure, and functionality as opposed to pure carbon-based graphene has led to a wide range of applications such as catalyst, energy and storage, biosensing, and nanoelectronics (Zhu, Du, and Lin. 2017). A recent study by Wang et al. synthesized 2D nanosheet GO/BiOCl *via* the hydrothermal method and observed a marked degradation for Rhodamine B up to 93.6% within 8 min. The study concluded that the excellent photocatalytic performance of the material was due to the large number of oxygen vacancies produced at the constrictive recombination interface, and the availability of channels for 2D-2D electron transport between GO and BiOCl (Wang et al. 2022).

2.2.2 BP

Black phosphorous (black P or BP), which was successfully synthesized for the first time in bulk form in 1914 by Bridgman, was recently revisited from the standpoint of a 2D layered material (Ling et al. 2015). Among the four major allotropes of phosphorus, BP is the least toxic and has the best thermodynamic stability (Li et al. 2019). BP is one of the metal-free direct bandgap layered semiconductors. Depending on the layer thickness, the BP bandgap can be adjusted from 2 to 0.3 eV (Sakthivel et al. 2020). BP demonstrates high carrier mobility ($\sim$1,000 cm^2 V^{-1} s^{-1}) phonon, optical properties due to its tunable bandgap, anisotropic electronic properties, and Brillouin Zone very near to the Fermi level. The orthorhombic layered structure of BP, in which the phosphorus (P) atoms are covalently bonded to three neighboring atoms, makes it possible to mechanically exfoliate thin films from bulk crystals (Zhang et al. 2019).

It is known that the photophysical processes in 2D semiconductors are significantly influenced by the Coulomb interactions (*e.g.* exciton dissociation) between excited electrons and holes (Guan et al. 2022). BP Quantum dots are known to have unique advantages during the charge separation process owing to their large specific surface with rich active sites and short carrier transport distance (Guan et al. 2022).

Numerous 2D BP-based nanomaterials have been prepared with excellent properties for various applications. For example, Co$_2$P/BP synthesized *via* the *in situ* growth method showed high selectivity for CO$_2$ reduction compared to bulk BP (Xu et al. 2022). The addition of metal phosphide, Co$_2$P to BP nanosheets, promoted the electron-hole separation process which enhanced the photocatalytic activity of the material.

2.2.3 *h*-BN

BN is a III–V compound with an equal number of B and N atoms. The most common crystal forms of BN in two stable phases are cubic (*c*-BN), and hexagonal (*h*-BN) while the metastable phases are wurtzite (*w*-BN) and rhombus (*r*-BN). Although *h*-BN exhibits an sp^2 bond structure extremely similar to that of graphene, it has a much wider bandgap of roughly 5.2 eV. *h*-BN, which is made up of regularly stacked planar networks of BN hexagons, and has a structure resembling graphite. Electrons tend to cluster around nitrogen atoms in h-BN giving it an insulating property, but in a graphene layer, the π electrons are delocalized, giving this material its semi-metallic nature (Lopes 2021). The construction of graphene devices with BN as the substrate can boost electron mobility because *h*-BN has a strong atomic affinity, a flat surface, and good thermal stability (Yang et al. 2022). Due to its high bandgap of around 6 eV (or 5.9 eV), BN is an electrical insulator as opposed to gapless graphene (Yang et al. 2022; Zhang et al. 2019). Boron and nitrogen atoms alternate to form BN nanosheets in a honeycomb-like lattice structure (Zhu, Du, and Lin 2017). *h*-BN is characterized by electrical insulation and high thermal conductivity up to 751 W mK^{-1} at 25 °C (Ren and Innocenzi 2021).

2.2.4 TMDs

Materials having the formula MX_2 are known as TMDs, where M stands for a transition metal element such as Mo, W, Hf, Ti, Zr, V, Nb, Ta, Re, *etc.* and X is an example of a chalcogen (S, Se, or Te) (Joshi et al. 2018).

TMDs with the structure form of X-M-X are made up of two layers of chalcogen atoms and a layer of transition metal atoms. All these three layers are joined through covalent bonds (Zhu, Du, and Lin 2017). The metal atoms in TMDs exhibit octahedral or trigonal prismatic coordination and their overall symmetry is hexagonal or rhombohedral (Wang et al. 2012).

Layered materials TMDs predominate from groups 4 (Ti, Zr, or Hf), 5, 6, or 7 (Te, Re), while TMDs from groups 8–10 are frequently found in non-layered structures. Each layer in layered structures typically has a thickness of 6~7 Å and is made up of a layer of metal atoms that are hexagonally packed and sandwiched between two layers of chalcogen atoms. The sandwich layers are connected by weak van der Waals forces, whereas the intralayer M-X bonds are primarily covalent, allowing the crystal to easily cleave along the layer surface (Chhowalla et al. 2013). TMDs have a variety of chemical and electrical properties as a result of the interaction between the 'd' orbitals from transition metals and 'p' orbitals from the chalcogenides (Jing et al. 2020). Their electronic band structure depends significantly on the number of layers in a single TMD material, particularly down to the single-layer limit (Mak et al. 2010).

The structures of TMDs are quite anisotropic. It uses a multi-layer structure. Each layer is joined by van der Waals forces, and the atoms inside the layers are connected by strong covalent bonds (Wang et al. 2012). Most TMDs have bandgap between 1.01 and 2.0 eV in the bulk and 1.31 to 2.38 eV in the monolayer (Wang et al. 2012). The bandgaps of the extensively researched semiconducting TMDs, such as MoS_2, $MoSe_2$, WS_2, and WSe_2, range from 1.09 to 1.57 eV in the bulk and from 1.55 to 1.99 eV in the monolayer (Guo and Robertson 2016).

2.2.5 LAYERED DOUBLE HYDROXIDES

The layered double hydroxide (LDH) materials are another class of 2D nanomaterials. These 2D structures have layered brucite-shaped structures with anionic clay networks (Wang, Xiang, and Li 2010). The formula used to define LDHs is $[M_{1-x2}{}^{+Mx}{}^{3+}(OH)_2]^{x+}[A^{n-}]_{x/n} \cdot mH_2O$, where M^{3+} and M^{2+} represent the trivalent and divalent cations, respectively. The A^{n-} indicates the intercalated anions, while 'm' denotes the number of H_2O molecules. The charge is indicated as the 'n' and the molar fraction of the trivalent cation is denoted in 'x' (Evans and Slade 2006).

The positive charge generated in an LDH nanosheet structure is by substituting some of the M^{+2} cations with M^{+3} cations in the brucite-shaped nanosheets. The balance in the total charges between the positive cations is attained by the presence of the intercalated anions that are present in between the hydrated layers of the nanosheets (Guo et al. 2010). There exist several methods of fabrication of these hydroxide nanomaterials such as the co-precipitation technique, ion exchange procedure, microwave light method, *etc.* These synthesis processes present their challenges and could be utilized based on the desired application (Wu et al. 2018).

One of the unique abilities showcased by this class of nanomaterials are the memory effect. In this process, the LDH nanomaterial loses its structure after the calcination process. However, the structure of this material is gained through the rehydration process (Misra and Perrotta 1992; V. Bugris et al. 2013; V. r. Bugris et al. 2013). LDHs are utilized in numerous applications for such as in the assembly of a dye-sensitized solar cell (George and Saravanakumar 2017), as catalyst precursor in catalytic applications (Sipos and Pálinkó 2018), as well as in environmental and medical fields (Laipan et al. 2020). The LDHs have also displayed potential for energy storage applications such as in supercapacitor studies in place of the electrode material (Guo et al. 2019).

2.2.6 MXENE

Another interesting class of nanomaterials is MXenes. It is a broad class of materials that is made up of 2D structures of transition metal carbides, carbonitrides, and nitrides (Naguib et al. 2011). The nanosheets are made up of early transition metals (*e.g.* V, Tr, Cr, Mo, W) with $n + 1$ ($n = 1,2,3$) layers (M layers). While the carbon and nitrogen atoms (the X atoms) are inserted in between these sheets (n layer). MXene is denoted using the formula of $M_{n+1}X_nT_x$, wherein T_x specifies the surface termination atoms (*e.g.* OH, O, F, Cl) which are connected to the external M layers. Owing to the range of nanomaterials synthesized in the class of MXene to date and the potential it offers due to the extensive range of their elements, this family of 2D materials is expanding promptly (Gogotsi and Anasori 2019). Contrary to the most compared 2D structure graphene, the MXene nanostructure could not be delaminated from the bulk material. These MXene sheets could be fabricated by growing them epitaxially on the substrate by several etching and delamination processes (Alhabeb et al. 2017; Anasori et al. 2015).

Some of the most commonly known MAX phase materials is titanium aluminum carbide (Ti_2AlC_2). This is the precursor or the first material used to prepare 2D sheets of titanium carbide (Alhabeb et al. 2017; Naguib et al. 2011). The nature of parent material often shows up in the product as well. This could be easily seen in the MXene sheets. The surface termination of these materials is the strongest property of this material which is available in their parent precursor as well. The surface termination contributes to the major electronic properties, creating them a strong option for electronic applications, for instance, metallic electrical conductors (Khazaei et al. 2016). The electronic attributes of MXene materials could be tweaked as a result of changing the surface termination nature and altering the M groups (Khazaei et al. 2013, 2016). Therefore, these 2D structures perform as metal, semiconductor, or insulator substances (Khazaei et al. 2016, 2017). These 2D nanomaterials are believed to be promising for a wide range of applications such as energy storage applications, energy harvesters, potentially used as electrode material and explored for various catalytic and sensors as well (Yan et al. 2017; Boota et al. 2016).

2.3 FABRICATION USING VAPOR-BASED GROWTH TECHNIQUE

2.3.1 TOP-DOWN APPROACH

Monolayer 2D nanomaterials are typically synthesized using two basic ways: top-down and bottom-up strategies. The development of 2D semiconductors with controllable architectural, physical, and chemical properties is crucial for many applications such as optoelectronics, sensing, catalysis, and energy storage and will open up new possibilities. However, most of them were produced in laboratories using very simple and straightforward mechanical exfoliation techniques from a piece of a naturally occurring crystalline sample, making them ineligible to be used in industrial applications due to lacking the ability to control the number of layers.

The top-down strategies typically employ chemical or mechanical exfoliation from bulk crystals to delaminate and exfoliate multi-layered materials. One or a few layers of 2D materials can be peeled off by mechanical force without disrupting the chemical bonding of the in-plane layer due to the weak van der Waals interaction between the layers of the materials. Novoselov et al. were the first to employ the method to create a graphene monolayer from highly oriented graphite (Novoselov et al. 2004). This technique is simple, fast, and low cost and the size of exfoliated 2D materials can be as small as tens of micrometers in size. As only shear force is used in the fabrication process and thus the fabricated nanosheets exhibited fewer defects compared to the bulk materials. It is commonly accepted that mechanically exfoliated 2D monolayers have superior quality and are clean on surfaces, making them ideal for basic research and the production of proof-of-concept devices. However, major limitations include the extremely low product yield and the unavoidable coexistence of thick flakes and nanosheets. Furthermore, due to shortcomings in terms of precision and reproducibility in hand-operated mechanical exfoliation technique is lacking, the size and geometry of the ultrathin 2D nanosheets are unstable and unregulated. Another top-down strategy is liquid exfoliation which can be categorized into sonication-assisted liquid exfoliation and shear force-assisted liquid exfoliation and has been largely reported to produce mono and few-layer 2D sheets such as MoS_2, WS_2, $MoSe_2$, h-BN, $NiTe_2$, Bi_2Te_3, and so on (Coleman et al. 2011). This technique uses bubbles created by sonication in the solution to wash bulk crystals that have been dispersed in a polar solvent and the resulting dispersion is then centrifuged to exfoliate layers in the solution. In this regard, a rapid burst of the sonication-generated bubbles could result in high energy, and restacking and aggregation between the 2D layers could be avoided provided the surface energies of the layered bulk materials and the liquid system are matched. In addition, important variables that affect the quantity and quality of 2D materials are the centrifugation rate and sonication time. The most popular solvents for ultrasonic exfoliation of MoS_2, WS_2, h-BN, and $MoTe_2$ include dimethylformamide (DMF), cyclohexyl-pyrrolidinone, N-methyl-pyrrolidone, N-dodecyl-pyrrolidone, ortho-dichlorobenzene, and isopropyl alcohol (Zhou et al. 2011).

2.3.2 Bottom-Up Approach (CVD)

Another option relies on the bottom-up assembly of inorganic, organic, or polymeric precursors in a 2D structure. One of the most useful and applicable synthesis techniques for the deposition of high-quality films is vapor-phase-based direct growth of ultrathin films. CVD, thermolysis, physical vapor transfer, and layer-by-layer conversion are examples of bottom-up synthetic processes. The two primary vapor-based deposition methods that offer distinct benefits for the large-area fabrication of nanostructured 2D semiconductors are CVD and atomic layer deposition (ALD).

CVD offers a scalable and controllable approach for the growth of high-quality 2D films. This technique involves activated chemical reactions of precursors as well as sequential reactions of gaseous components in a vapor state on the surface of substrates in a precisely tailored environment. The interfacial properties, geometrical characteristics of the substrate, and structural phases of solid films all have a significant impact on the properties of 2D nanomaterials. Generally, CVD growth is affected by mass transfer, heat transfer and interfacial reactions on the surface and could be controlled by manipulating the deposition parameters including temperature, pressure, substrate, and precursors where they are highly dependent on the structural phases, morphology, and interfacial reactions. For example, Somani et al. published the first study on the formation of few-layer graphene on the polycrystalline Ni film *via* CVD in 2006 (Somani, Somani, and Umeno 2006). The graphene monolayer is then produced using the CVD method on a Ni film that has been formed on a SiO_2/Si substrate (Pollard et al. 2009). The Ni film served as both the catalyst to induce the formation of precursors to growing graphene sheets as well as a substrate to aid in the synthesis of graphene. Notably, the precursors, Ni film deposited on SiO_2/Si substrate, temperature, catalysts, and atmospheres are key factors affecting the structural property of graphene.

2.3.2.1 Temperature

One of the most crucial fundamental parameters during CVD growth is temperature, and it directly impacts several interconnected variables such as the flow rate of carrier gas and film growth rate. In general, the excellent quality of CVD films can be deposited when the temperature is high. In general, high temperatures allow for the deposition of CVD films of excellent quality. However, extremely high temperatures do have downsides, especially when they can result in a large concentration gradient that makes the mass transfer and flow in the system unstable. In 2D TMD nanostructures, for instance, high gas pressure and concentration allowed for the manageable synthesis of MoS_2, WS_2, and other compounds. The thermodynamically triggered deposition method is typically facilitated by high CVD temperatures and appropriate pressure, whereas low CVD temperatures typically result in the kinetic growth process.

2.3.2.2 Pressure

The pressure that is employed in CVD can vary from ambient pressure to a few millimeters. Low pressure is always more favorable for the uniform growth of ultrathin 2D nanostructures at the wafer-scale deposition of TMD 2D nanostructures. This is due to the best conditions for a controllable CVD process reached at the low concentration and high velocity of the mass feed. Additionally, the consistent layer-by-layer growth of TMD 2D films can be directly impacted by the partial pressure of components. For instance, when growing a 2D MoS_2 film by CVD, the fabrication of the second fundamental layer over the previously developed film can only begin near the grain boundaries of the initially deposited layer at low pressure.

2.3.2.3 Substrates

The formation of continuous layers on the substrate and the quantity of layers deposited are significantly influenced by the quality of the metal surface. For the formation of monolayer and few-layer h-BN by CVD, for instance, poly- or mono-crystalline Ni and Cu are frequently used as substrates. The h-BN growth's nucleation is influenced by the shape of the substrate's surface. In a different instance, the treatment of the substrate prior to the growth has a significant impact on the growth of MoS_2. The lateral development of MoS_2 is promoted by the substrate treatment with aromatic molecules like reduced graphene oxide (r-GO), and it is predicted that the presence of aromatic molecules will improve the growth surface's wetting and reduce the free energy for nucleation (Lee et al. 2012). In contrast, no monolayer MoS_2 can be observed on the substrates following the growth when inorganic seeding promoters such as aluminum oxide, hafnium oxide, and silica are used.

2.3.2.3.1 Atomic Layer Deposition

ALD can be seen as a modification of CVD and various strategies can be employed for the thickness-controlled fabrication of sandwiched heterostructures and 2D semi-conductors. The main difference between ALD and other vapor deposition techniques is that after each precursor exposure, a self-saturating surface monolayer is created. ALD uses a sequence of self-limiting, surface-saturated reactions as a deposition process to create thin conformal films at a controlled rate. Fabrication of defects-free 2D nanofilms and their heterostructures on the wafer scale with accurate control of thickness is only possible with ALD. For instance, the synthesis of WSe_2 utilizing WCl6 and diethyl selenide (DESe) at the higher ALD temperature of 600–800 °C was carried out using a self-limiting technique, and the dependence of film thickness on the ALD processing temperature was noted. On the Si/SiO_2 wafers, five, three, and one crystallized layer of WSe_2 were deposited at 600, 700, and 800 °C, respectively (Browning et al. 2016). On the other hand, WF_6 and H_2S are used as precursors to deposit layered 2D WS_2 at 300 °C; where a Zn-based catalyst is required to initiate the deposition of WS_2 film (Delabie et al. 2015).

2.4 SOLUTION-BASED SYNTHESIS METHODS AND CHEMICAL FUNCTIONALIZATION

The use of wet chemical techniques for the synthesis of 2D nanomaterials is a game-changing process. The use of precursor materials and the ability to tune the necessary physical factors results in even isolating nanomaterials at varying thickness levels from monolayers to multiple layers. This process also provides the ability to dope or introduce stoichiometric alterations to improve the chemical or physical properties. Moreover, the large-scale synthesis process benefits from the use of a solution-based synthesis protocol compared to the use of tedious processes such as CVD. Apart from that, the choice of precursor material can also enable the funda-mental applying a sustainable manufacturing process. In this section, we discuss a few of the solution-based synthesis processes to produce 2D nanomaterials.

2.4.1 Hydrothermal/Solvothermal Synthesis

One of the widely reported techniques that opted to produce a wide variety of TMDs and various other 2D nanomaterials is the hydrothermal/solvothermal technique

(Ganguly, Mathew, et al. 2019). A solvothermal or hydrothermal synthesis is a process of precursor reactants being allowed to be sealed and heated at pressurized atmosphere which results in an accelerated chemical reaction. When the choice of solvent used in the reaction is water, then the process is named as hydrothermal, while the use of any other solvent makes the reaction termed as a solvothermal technique. The hydrothermal/solvothermal synthesis process is different from any solid-state reaction. In any solid-state process, the reaction occurs at the interface; on the other hand, the hydrothermal/solvothermal process occurs at the atomic level and that too in the solution phase (Ganguly et al. 2019). The decrease in the operating temperature and the reduction of the overall complexity of this technique make it a productive option for synthesizing several classes of novel materials. In a chapter, Feng et al. discussed a very thorough knowledge of the hydrothermal method. The authors emphasize the significance of solvent choice in the synthesis procedure. Some of the vital factors such as density, surface tension, and viscosity definitively affect the kinetics of the reaction. This synthesis is accomplished using a high-pressure vessel known as the autoclave (Figure 2.3). These autoclaves or hydrothermal reactors are appropriate for working below intense temperatures and pressure. These vessels are tolerant against acid, alkali, and other oxidants. They have a distinctive sealing aspect which aids them to maintain the heat and pressure for an extended period of the experimental procedure. There have been abundant articles and an entire section devoted to applying this production method in nanomaterial synthesis. Articles of solvothermal or hydrothermal synthesis for 2D nanomaterials and composites have also been widely reported (Han et al. 2017; Yu et al. 2017). The uniformity of the nanomaterial prepared and the tuning of the shape and size of the product makes this process unique (Feng and Li 2017).

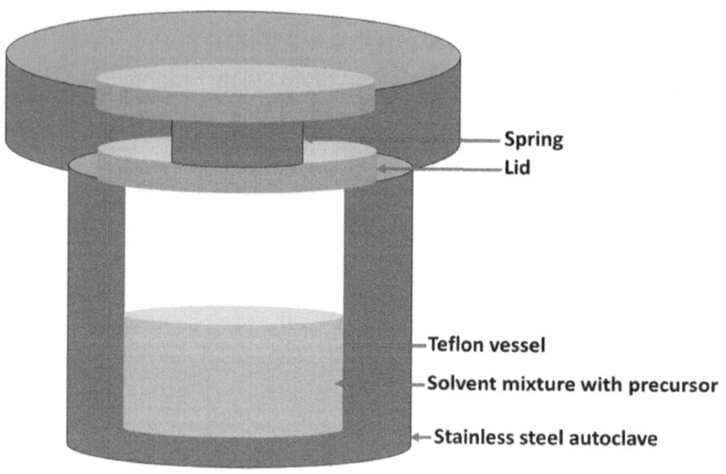

FIGURE 2.3 Schematic illustration of an autoclave vessel used for hydrothermal or solvothermal synthesis. Reproduced with permission of ref (Ganguly, Harb, et al. 2019).

2.4.2 MICROWAVE-ASSISTED SYNTHESIS

Microwave-assisted synthesis is an interesting technique that utilizes microwave energy to synthesize nanomaterials. This synthesis process can be carried out in open or sealed vessels. Moreover, it can be performed at atmospheric pressure using a wide range of organic/inorganic solvents. The choice of solvent plays a crucial role in controlling the reaction kinetics and determining the shape and size of the nanomaterials prepared (Roy et al. 2020). This constraint is very similar to the solvothermal technique, which we have discussed in detail. The heating process in the microwave-assisted synthesis process is quite quick compared to conventional heating techniques. Several reports highlight the use of microwave energy to synthesize various 2D nanomaterials (Bajpai and Wagner 2015). Multi-walled carbon nanotubes, and several transition metal chalcogenides are a few among the growing list of nanomaterials prepared using this method. Tungsten disulfide (WS_2) nanowires were prepared using ethanolamine as the solvent and tungstic acid and sulfur as the precursor. The solution was microwaved for 5–6 min and then acidified to obtain a precipitate. The precipitate was further calcined at 750 °C for 90 min in an inert (Ar) atmosphere (Panigrahi and Pathak 2008). Similarly, nanosheets of molybdenum disulfide with interesting morphological details were prepared using DMF as a solvent (Gao, Chan, and Sun 2015). Heating ammonium molybdate in the solvent inside a closed microwavable vessel at 240 °C for 120 min resulted in edge terminated and expanded interlayer spaced sheets of MoS_2 (Figure 2.4). Similar reports of producing this film structure of MoS_2, WS_2, $MoSe_2$, and WSe_2 using microwave irradiation at a time interval of 45–90 sec exist in literature (Giri et al. 2017). The thin films were grown on a solid substrate and the precursors were solubilized in 1-butyl 3-methylimidazolium tetrafluoroborate solvent which was further sandwiched between two substrates. The use of a combination of microwave and other techniques such as solvothermal or exfoliation is also an

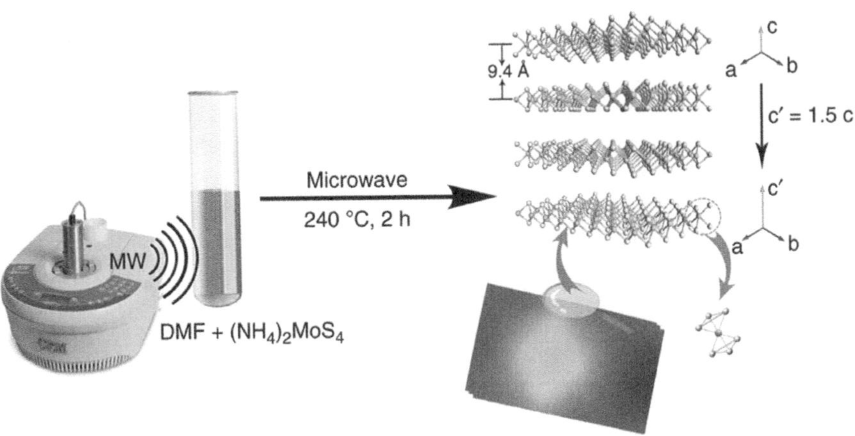

FIGURE 2.4 Schematic illustration of edge terminated and interlayer spacing enhanced MoS_2 sheets using microwave synthesis. Reproduced with permission of ref (Gao, Chan, and Sun 2015).

interesting method that has been explored in some of the recent reports. The growth of MoS$_2$ film on a gold substrate was carried out using microwave-assisted solvothermal technique. The synthesis was carried out at 200 °C under a nitrogen atmosphere (Lee et al. 2019). In another report, spherical and rod-like structures of MoS$_2$ nanomaterials were obtained through microwave irradiation, followed by the exfoliation of the MoS$_2$ nanomaterial in a liquid medium at 175 °C for 10 min (Reshmi et al. 2017). More recently, reports of the synthesis of single-walled carbon nanotubes using microwave-assisted methods were also found to be quite interesting and promising. Amidst the growth of this process, there lie a few drawbacks such as the use of solvents for the microwave energy to be dispersed thoroughly across the reaction media. This is limited as the reaction process is skewed under the boiling point of the solvent. Most of the inorganic/organic solvents used have a maximum temperature range of less than 300 °C. Synthesis at lower temperatures often results in the growth phase of nanomaterials with lower crystallinity.

2.4.3 TEMPLATE SYNTHESIS

Template synthesis is one of the most effective bottom-up approaches to synthesizing 2D nanomaterials. In general, templates induce gaseous molecules or liquid phase ions to preferentially grow toward chosen orientation. This is effectively aided by the various surface activities and their dimensions. Crucial factors influencing the growth such as the adsorption energy barrier, catalytic activity, and hydrophilic property might determine the progress of nanomaterial synthesis. In the case of 2D nanomaterials, the growth along the z-axes is constrained. Moreover, the chosen template used in the synthesis process refrains from participation in the chemical reaction and they remain quite stable (Liu, Goebl, and Yin 2013). Apart from these features of stability, the dimension of the template also contributes toward the production of the desired dimensional nanostructure. Hence, for synthesizing a 2D nanostructure, the use of a template material with lower dimensions such as quantum dots and 1D nanowires is futile (Xiao et al. 2016).

Some of the reports on template-based synthesis include an earlier article by Sun et al. They reported a synthesis process of GO and titania hybrid laminate (GO/TiO$_2$) which utilizes GO monolayer as a universal template. The synthesized laminated hybrid displayed excellent water desalination application. Following it, in another report, it was established that the distance between the GO layers could aid in fabricating various other layered structures of metal oxides (Sun et al. 2015). Considering this property, polycrystalline TiO$_2$, ZnO, Fe$_2$O$_3$ and amorphous SiO$_2$ were effectively deposited on the GO layers. The deposited layers displayed a thickness of several nanometers. Moreover, various textured structures of graphene sheets such as wrinkled, crumpled could be also potentially used as a template to fabricate various architectural structures of crystalline layer of metal oxides (Figure 2.5). 2D metal oxides nanomaterials such as ZnO, CuO, Mn$_2$O$_3$, and Al$_2$O$_3$ were synthesized in this process (Saito et al. 2015). The metal ions were intercalated between the van der Waals gaps of GOs (Chen et al. 2016). This is subsequently results *in situ* conversion and annealing, where the ultrathin 2D metal oxide nanoplates were obtained. The synthesis process is subdivided into two fundamental aspects; (a) the presence of strong

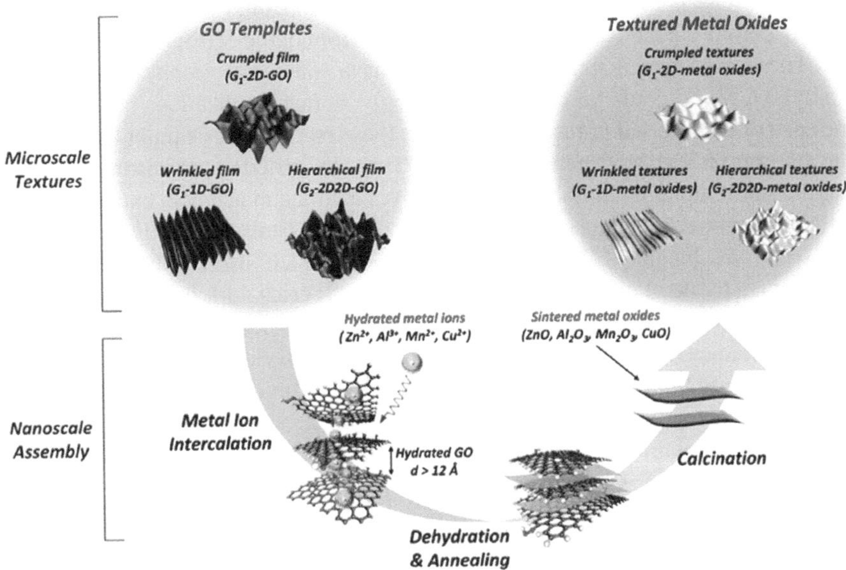

FIGURE 2.5 Formation of hierarchical metal oxide architectures using textured GO as a versatile ion intercalation template. Reproduced with permission of ref (Chen et al. 2016).

electrostatic attraction between the metal ions and the GO sheets which forms strong adsorption and (b) the chemical complex formation; the carboxylic groups present in the GO sheets have high binding energy with the metal ions and thus favoring the formation of chemically specific ion-carboxylate complexation. Even though there have been several reports of graphene and its derivatives used as templates, there exist several challenges which cannot be undermined. Stacking issues between the graphene layers frequently affect the development of 2D nanosheets. Moreover, the extra steps taken to remove the template such as calcination could potentially impact the surface properties of the products attained.

2.4.4 TOPOCHEMICAL SYNTHESIS

An alternative approach for synthesis of novel 2D non-layered structured materials, that has fascinated attention lately is the salt-templating method and topochemical transformation technique. In the previous section, we discussed templating techniques, and, in this section, we discuss another interesting approach. This combination could be potentially used for several non-layered materials, such as 2D h-MoO₃ (Nicolosi et al. 2013). The topochemical synthesis process offers alternate methods of producing 2D materials which do not necessitate van der Waals bonded solid precursors or vapor phase deposition. The crystal structures of non-layered nanomaterials have strong long range chemical bonds across all dimensions. This essentially contributes to being a hindrance to breaking the bonds between them (Hanlon et al. 2014). Thus, the lack of intrinsic driving force for anisotropic growth for 2D nanostructure is a challenge for non-layered precursors. 2D transition metal oxides (TMOs) with unique electronic properties and high specific surface areas have displayed potential applications stretching from optoelectronics to energy storage. These non-layered structured materials have some exceptional properties. For example, h-MoO$_3$ could be exploited for energy storage applications because of the presence of big internal pores. Nevertheless, these metal oxides cannot be exfoliated into 2D morphology by mere exfoliation. Thus, by utilizing interesting processes like the salt-templating method, which involves a lattice match concept to transform a precursor into a 2D material on the surface of salt crystals. Similarly, several other nanostructures include TMDs and metal carbides which also can be topochemical transformed to metal nitrides. Such understanding is crucial for underlying concepts as the synthesis products could be compared and studied for stoichiometric and chemical studies. For example, a topochemical process yielded nitrides of MXenes from carbide MXenes which showed enhanced electronic conductivity for nitrides compared to their carbide start material.

The salt-templating strategy was reported recently for the synthesis of many oxides as shown in Figure 2.6. Usually, ethanol as a solvent is used to dissolve a metal precursor to make a solution. This is further mixed with NaCl or KCl salt. This is labeled as precursor@salt in the schematic given below. Allowing the salt precursors to dry and then annealing them at high temperatures results in the formation of 2D oxides@salt (Xiao et al. 2016). The 2D oxide nanomaterials have finally been developed on the surface of the salt crystals, where this salt could be further removed by dissolving the obtained product in deionized water. The remaining product which remains undissolved in the water is the 2D metal oxide sheets which could be obtained by vacuum filtration. In another instance, the lattice matching process of the crystal face of the salt and the required metal oxide was further used in this case with NaCl. Here, the microcrystals of the salt acted as the substrate/template to guide the oxide growth at increased temperatures. The formation of h-MoO$_3$ on the surface of NaCl was interesting as there could be other forms of the same oxides (*e.g.* α-MO$_3$) which could be possibly developed. This is negated not just by the lattice mismatch but also due to the presence of sodium ions (Na$^+$) which serves as a mineralizer and stabilizer for the hexagonal tunnels of h-MoO$_3$. This synthesis route highlights the novel and facile fabrication process which has been expanded to other 2D TMOs. In recent reports, Zavabeti et al. reported a liquid metal-based reaction process. This utilized metals which are in a liquid state at room temperature as a reaction environment to synthesize other 2D TMOs such as HfO$_2$, Al$_2$O$_3$, and Gd$_2$O$_3$. The formation of 2D structures for these nanomaterials was unimaginable previously (Zavabeti et al. 2017). Exposure of air on the surface of the liquid metal droplet could result in the formation of nanosheets of the 2D oxide materials. Later, the oxide layer could be deposited on the substrate from the parent liquid droplet by merely touching the substrate. Devoid of any strong interaction between the layer of oxide and the metal makes this transfer quite precise and efficient. Another alternative method involves the insertion of high-pressure air into the liquid metal. On passing, water bubbles passing through the parent liquid metal could result in dispersing the oxide sheets into the aqueous solution. This technique has the potential to be scalable and results

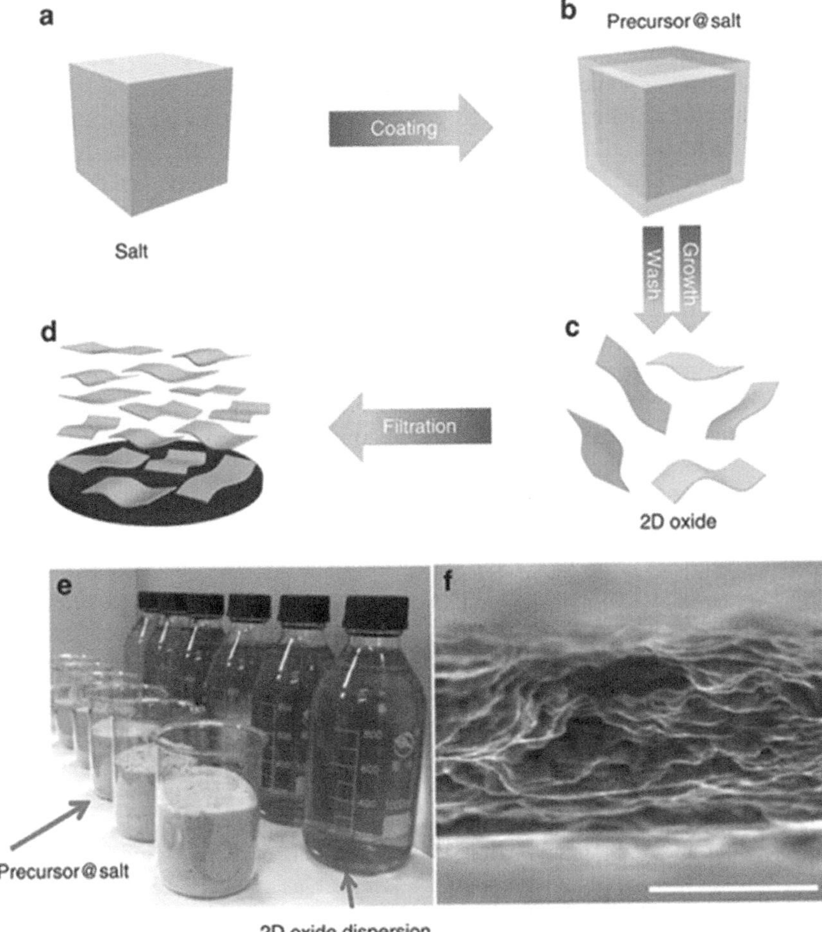

FIGURE 2.6 (a, b) Schematic representation of the fabrication process. The precursor solution was coated on the salt by mixing the solution with a large number of salt microcrystals. (c) After thermal annealing and rinsing in water, 2D oxide flakes were obtained (d).

in the formation of a high yield of desired 2D MOs. This process has a promising applicability to be extended to several other classes of nanomaterials (Chandler, Weeks, and Andersen 1983).

Pseudocapacitor electrodes were fabricated by filtering the 2D oxide/ethanol dispersion using a Celgard separator. (e) Optical image of the precursors and 2D h-MoO$_3$-ethonal dispersions. The front row shows six beakers containing the precursor, and the back row shows six bottles containing the 2D h-MoO$_3$-ethanol dispersions. (f) Multi-layered 2D oxide is shown in the cross-sectional SEM image. Scale bar, 1 µm. Reproduced with permission of ref (Xiao et al. 2016).

2.5   DISCUSSION AND OUTLOOK

This chapter highlights various methods employed to produce 2D nanomaterials. In many instances, some methods use a combination of more than one technique to achieve the best structural output. As explained in the previous section, a microwave synthesis technique employed with a post-calcination can effectively improve the structure of the 2D nanomaterial with high surface area and increased defects, which in turn plays an active role in catalysis and energy storage functions. The use of such hybrid methods could be the future way toward making nanomaterials with unique structural and chemical properties.

There is no doubt to assume the expansion of the 2D family, as we head toward the upcoming decades equipped with newer and better synthesis routes such as the template and topochemical synthesis. Materials such as metal oxides bonded with strong ionic bonds were never assumed to be reduced in 2D structures. However, these synthesis methods have made those assumptions to be untrue and several such reports have been explained in the previous sections.

The growth of optical imaging support with advanced scanning electron microscopes, transmission electron microscopes, atomic force microscopes, *etc.* cannot be neglected while we only talk about the synthesis protocols. These imaging microscopes have enabled us to probe these materials at their lowest possible dimensions, which was never accessible before. Similarly, other advanced theoretical modeling software has enabled to predict the behavior and the activity of 2D nanomaterials with newer defects and more so with chemical functionalization. As we head toward more advanced computational advancements, it is likely to see those spillovers to the world of materials. The use of artificial intelligence could see the game-changing the output to the preferred synthesis of newer structures and their applications.

Apart from all these highlights and glamour of these nanomaterials, there remain many fundamental challenges that need redressal. Structural modifications do improve in terms of applications point of view such as the increase in catalytic sites. However, the overall stability of the material might be compromised in this process. Moreover, an increase in catalytic sites does improve catalytic reactions and improve the overall performance for the desired application. However, there may be instances where these sites do promote side reactions which in turn can produce impurities and can cause damage to the structure. Similarly, the expansion of the interlayer would lower the ion diffusion process in the materials, but it equally opens up challenges to the stability of the overall structure of the material. The selective trade-offs between the stability and the changes introduced, whether it is physical or chemical to achieve a better result is quite an interesting challenge and question posed for future research. Apart from these, fundamental understanding of the mechanistic transfer of electrons within the structures, engineered defects, and their overall impact needs to be carefully studied to exploit their application in catalysis, medicinal nanoscience, energy storage, and much further. Theoretical estimation and better modeling tools can assist in this venture and provide better answers to find a synthesis process that could result in materials with interesting applications. Moreover, scalability, commercialization, and sustainable choices should be at the heart of decision-making. As we head into the new decade, compromising on environmental standards is not an option. Therefore, the stakes are quite high to pursue such new methods.

2.6 CONCLUSIONS

This chapter provides a glimpse of the different techniques used to produce different types of 2D nanomaterials. The sections discuss the interesting features of these techniques and also entail their limitations. The use of hybrid synthesizing processes has been a highlight among them. The chapter also discusses the high stakes involved in finding commercially viable, scalable, environmentally friendly, and sustainable methods to synthesize nanomaterials.

REFERENCES

Alhabeb, M., K. Maleski, B. Anasori, P. Lelyukh, L. Clark, S. Sin, and Y. Gogotsi. 2017. Guidelines for synthesis and processing of two-dimensional titanium carbide (Ti$_3$C$_2$T x MXene). *Chemistry of Materials* 29 (18):7633–7644.

Anasori, B., Y. Xie, M. Beidaghi, J. Lu, B. C. Hosler, L. Hultman, and M. W. Barsoum. 2015. Two-dimensional, ordered, double transition metals carbides (MXenes). *ACS Nano* 9 (10):9507–9516.

Bajpai, R., and H. D. Wagner. 2015. Fast growth of carbon nanotubes using a microwave oven. *Carbon* 82:327–336.

Boota, M., B. Anasori, C. Voigt, M. Q. Zhao, M. W. Barsoum, and Y. Gogotsi. 2016. Pseudocapacitive electrodes produced by oxidant-free polymerization of pyrrole between the layers of 2D titanium carbide (MXene). *Advanced Materials* 28 (7):1517–1522.

Browning, R., N. Kuperman, R. Solanki, V. Kanzyuba, and S. Rouvimov. 2016. Large area growth of layered WSe2 films. *Semiconductor Science and Technology* 31 (9):095002.

Bugris, V., H. Haspel, A. Kukovecz, Z. Kónya, M. Sipiczki, P. Sipos, and I. Pálinkó. 2013. Rehydration of dehydrated CaFe-L (ayered) D (ouble) H (ydroxide) followed by thermogravimetry, X-ray diffractometry and dielectric relaxation spectroscopy. *Journal of Molecular Structure* 1044:26–31.

Bugris, V. r., H. Haspel, A. k. Kukovecz, Z. n. Kónya, M. N. Sipiczki, P. l. Sipos, and I. n. Pálinkó. 2013. Water types and their relaxation behavior in partially rehydrated CaFe-mixed binary oxide obtained from CaFe-layered double hydroxide in the 155–298 K temperature range. *Langmuir* 29 (43):13315–13321.

Chandler, D., J. D. Weeks, and H. C. Andersen. 1983. Van der Waals picture of liquids, solids, and phase transformations. *Science* 220 (4599):787–794.

Chen, C., J. Hu, X. Yang, T. Yang, J. Qu, C. Guo, and C. M. Li. 2021. Ambient-stable black phosphorus-based 2D/2D S-scheme heterojunction for efficient photocatalytic CO^2 reduction to syngas. *ACS Applied Materials & Interfaces* 13 (17):20162–20173.

Chen, P.-Y., M. Liu, T. M. Valentin, Z. Wang, R. Spitz Steinberg, J. Sodhi, and R. H. Hurt. 2016. Hierarchical metal oxide topographies replicated from highly textured graphene oxide by intercalation templating. *ACS Nano* 10 (12):10869–10879.

Chhowalla, M., H. S. Shin, G. Eda, L.-J. Li, K. P. Loh, and H. Zhang. 2013. The chemistry of two-dimensional layered transition metal dichalcogenide nanosheets. *Nature Chemistry* 5 (4):263.

Coleman, J. N., M. Lotya, A. O'Neill, S. D. Bergin, P. J. King, U. Khan, and R. J. Smith. 2011. Two-dimensional nanosheets produced by liquid exfoliation of layered materials. *Science* 331 (6017):568–571.

Delabie, A., M. Caymax, B. Groven, M. Heyne, K. Haesevoets, J. Meersschaut, and P. Verdonck. 2015. Low temperature deposition of 2D WS 2 layers from WF 6 and H 2 S precursors: impact of reducing agents. *Chemical Communications* 51 (86):15692–15695.

Evans, D. G., and R. C. Slade. 2006. Structural aspects of layered double hydroxides. *Layered Double Hydroxides* 119:1–87.

Fatima, J., A. N. Shah, M. B. Tahir, T. Mehmood, A. A. Shah, M. Tanveer, and S. Alansi. 2022. Tunable 2D nanomaterials; their key roles and mechanisms in water purification and monitoring. *Frontiers in Environmental Science* 10:1–23, Article .766743

Feng, S.-H., and G.-H. Li. 2017. Hydrothermal and solvothermal syntheses. In *Modern Inorganic Synthetic Chemistry* (Second Edition, pp. 73–104), edited by R. Xu, and Y. Xu: Elsevier.

Ganguly, P., M. Harb, Z. Cao, L. Cavallo, A. Breen, S. Dervin, and S. C. Pillai. 2019. 2D nanomaterials for photocatalytic hydrogen production. *ACS Energy Letters* 4 (7):1687–1709.

Ganguly, P., S. Mathew, L. Clarizia, R. S. Kumar, A. Akande, S. J. Hinder, and S. C. Pillai. 2019. Ternary metal chalcogenide heterostructure (AgInS$_2$–TiO$_2$) nanocomposites for visible light photocatalytic applications. *ACS Omega* 5 (1):406–421.

Ganguly, P., S. Mathew, L. Clarizia, S. R. Kumar, A. Akande, S. Hinder, and C. S. Pillai. 2019. Theoretical and experimental investigation of visible light responsive AgBiS$_2$ -TiO$_2$ heterojunctions for enhanced photocatalytic applications. *Applied Catalysis B: Environmental* 253:401–441. https://doi.org/10.1016/j.apcatb.2019.04.033

Gao, M.-R., M. K. Chan, and Y. Sun. 2015. Edge-terminated molybdenum disulfide with a 9.4-Å interlayer spacing for electrochemical hydrogen production. *Nature Communications* 6 (1):1–8.

Geim, A. K., and I. V. Grigorieva. 2013. Van der Waals heterostructures. *Nature* 499 (7459):419–425.

George, G., and M. Saravanakumar. 2017. Synthesising methods of layered double hydroxides and its use in the fabrication of dye sensitised solar cell (DSSC): a short review. *IOP Conference Series: Materials Science and Engineering* 263 (3):032020.

Giri, A., H. Yang, K. Thiyagarajan, W. Jang, J. M. Myoung, R. Singh, and U. Jeong. 2017. One-step solution phase growth of transition metal dichalcogenide thin films directly on solid substrates. *Advanced Materials* 29 (26):1700291.

Gogotsi, Y., and B. Anasori. 2019. The rise of MXenes. *ACS Nano* 13:8491–8494.

Guan, R., L. Wang, D. Wang, K. Li, H. Tan, Y. Chen, and Z. Sun. 2022. Boosting photocatalytic hydrogen production via enhanced exciton dissociation in black phosphorus quantum Dots/TiO$_2$ heterojunction. *Chemical Engineering Journal* 435:135138.

Guo, D., X. Song, L. Tan, H. Ma, W. Sun, H. Pang, and X. Wang. 2019. A facile dissolved and reassembled strategy towards sandwich-like rGO@ NiCoAl-LDHs with excellent supercapacitor performance. *Chemical Engineering Journal* 356:955–963.

Guo, X., F. Zhang, D. G. Evans, and X. Duan. 2010. Layered double hydroxide films: synthesis, properties and applications. *Chemical Communications* 46 (29):5197–5210.

Guo, Y., and J. Robertson. 2016. Band engineering in transition metal dichalcogenides: stacked versus lateral heterostructures. *Applied Physics Letters* 108 (23):233104.

Han, C., Y. Zhang, P. Gao, S. Chen, X. Liu, Y. Mi, and J. Chang. 2017. High-yield production of MoS$_2$ and WS$_2$ quantum sheets from their bulk materials. *Nano Letters* 17 (12):7767–7772.

Hanlon, D., C. Backes, T. M. Higgins, M. Hughes, A. O'Neill, P. King, and H. Pettersson. 2014. Production of molybdenum trioxide nanosheets by liquid exfoliation and their application in high-performance supercapacitors. *Chemistry of Materials* 26 (4):1751–1763.

Hu, T., X. Mei, Y. Wang, X. Weng, R. Liang, and M. Wei. 2019. Two-dimensional nanomaterials: fascinating materials in biomedical field. *Science Bulletin* 64 (22):1707–1727.

Jing, Y., B. Liu, X. Zhu, F. Ouyang, J. Sun, and Y. Zhou. 2020. Tunable electronic structure of two-dimensional transition metal chalcogenides for optoelectronic applications. *Nanophotonics* 9 (7):1675–1694.

Joshi, N., T. Hayasaka, Y. Liu, H. Liu, O. N. Oliveira, and L. Lin. 2018. A review on chemiresistive room temperature gas sensors based on metal oxide nanostructures, graphene and 2D transition metal dichalcogenides. *Microchimica Acta* 185 (4):1–16.

Khazaei, M., M. Arai, T. Sasaki, C. Y. Chung, N. S. Venkataramanan, M. Estili, and Y. Kawazoe. 2013. Novel electronic and magnetic properties of two-dimensional transition metal carbides and nitrides. *Advanced Functional Materials* 23 (17):2185–2192.

Khazaei, M., A. Ranjbar, M. Arai, and S. Yunoki. 2016. Topological insulators in the ordered double transition metals M 2′ M ″C 2 MXenes (M′= Mo, W; M ″= Ti, Zr, Hf). *Physical Review B* 94 (12):125152.

Khazaei, M., A. Ranjbar, M. Arai, T. Sasaki, and S. Yunoki. 2017. Electronic properties and applications of MXenes: a theoretical review. *Journal of Materials Chemistry C* 5 (10):2488–2503.

Laipan, M., J. Yu, R. Zhu, J. Zhu, A. T. Smith, H. He, and L. Sun. 2020. Functionalized layered double hydroxides for innovative applications. *Materials Horizons* 7 (3):715–745.

Lee, J. Y., J.-H. Shin, G.-H. Lee, and C.-H. Lee. 2016. Two-dimensional semiconductor optoelectronics based on van der Waals heterostructures. *Nanomaterials* 6 (11):193.

Lee, Y. B., S. K. Kim, S. Ji, W. Song, H.-S. Chung, M. K. Choi, and K.-S. An. 2019. Facile microwave assisted synthesis of vastly edge exposed 1T/2H-MoS$_2$ with enhanced activity for hydrogen evolution catalysis. *Journal of Materials Chemistry A* 7 (8):3563–3569.

Lee, Y. H., X. Q. Zhang, W. Zhang, M. T. Chang, C. T. Lin, K. D. Chang, and L. J. Li. 2012. Synthesis of large-area MoS2 atomic layers with chemical vapor deposition. *Advanced Materials* 24 (17):2320–2325.

Li, B., C. Lai, G. Zeng, D. Huang, L. Qin, M. Zhang, and C. Zhou. 2019. Black phosphorus, a rising star 2D nanomaterial in the post-graphene era: synthesis, properties, modifications, and photocatalysis applications. *Small* 15 (8):1804565.

Ling, X., H. Wang, S. Huang, F. Xia, and M. S. Dresselhaus. 2015. The renaissance of black phosphorus. *Proceedings of the National Academy of Sciences* 112 (15):4523–4530.

Liu, Y., J. Goebl, and Y. Yin. 2013. Templated synthesis of nanostructured materials. *Chemical Society Reviews* 42 (7):2610–2653.

Lopes, J. M. J. 2021. Synthesis of hexagonal boron nitride: from bulk crystals to atomically thin films. *Progress in Crystal Growth and Characterization of Materials* 67 (2):100522.

Mak, K., C. Lee, J. Hone, J. Shan, and T. Heinz. 2010. Atomically thin MoS$_2$: a new direct-gap semiconductor [Article]. *Physical Review Letters* 105 (13), Article ARTN 136805. https://doi.org/10.1103/PhysRevLett.105.136805

Misra, C., and A. Perrotta. 1992. Composition and properties of synthetic hydrotalcites. *Clays and Clay Minerals* 40 (2):145–150.

Naguib, M., M. Kurtoglu, V. Presser, J. Lu, J. Niu, M. Heon, and M. W. Barsoum. 2011. Two-dimensional nanocrystals produced by exfoliation of Ti$_3$AlC$_2$. *Advanced Materials* 23 (37):4248–4253.

Nicolosi, V., M. Chhowalla, M. Kanatzidis, M. Strano, and J. Coleman. 2013. Liquid exfoliation of layered materials [Review]. *Science* 340 (6139):1420–+, Article ARTN 1226419. https://doi.org/10.1126/science.1226419

Novoselov, K. S., A. K. Geim, S. V. Morozov, D.-e. Jiang, Y. Zhang, S. V. Dubonos, and A. A. Firsov. 2004. Electric field effect in atomically thin carbon films. *Science* 306 (5696):666–669.

Panigrahi, P. K., and A. Pathak. 2008. Microwave-assisted synthesis of WS$_2$ nanowires through tetrathiotungstate precursors. *Science and Technology of Advanced Materials* 9 (4):045008.

Pillai, S. C., and P. Ganguly. 2021. 2D nanomaterials and composites for energy storage and conversion. In *2D Materials for Energy Storage and Conversion*, edited by Suresh C. Pillai: IOP Publishing.

Pollard, A., R. Nair, S. Sabki, C. Staddon, L. Perdigao, C. Hsu, and A. Geim. 2009. Formation of monolayer graphene by annealing sacrificial nickel thin films. *The Journal of Physical Chemistry C* 113 (38):16565–16567.

Ren, J., and P. Innocenzi. 2021. 2D Boron nitride heterostructures: recent advances and future challenges. *Small Structures* 2 (11):2100068.

Reshmi, S., M. Akshaya, B. Satpati, A. Roy, P. K. Basu, and K. Bhattacharjee. 2017. Tailored MoS$_2$ nanorods: a simple microwave assisted synthesis. *Materials Research Express* 4 (11):115012.

Roy, S., R. Bajpai, R. P. Biro, and H. D. Wagner. 2020. Fast growth of nanodiamond in a microwave oven under atmospheric conditions. *Journal of Materials Science* 55 (2):535–544.

Saito, Y., X. Luo, C. Zhao, W. Pan, C. Chen, J. Gong, and H. Wu. 2015. Filling the gaps between graphene oxide: a general strategy toward nanolayered oxides. *Advanced Functional Materials* 25 (35):5683–5690.

Sakthivel, T., X. Huang, Y. Wu, and S. Rtimi. 2020. Recent progress in black phosphorus nanostructures as environmental photocatalysts. *Chemical Engineering Journal* 379:122297.

Sipos, P., and I. Pálinkó. 2018. As-prepared and intercalated layered double hydroxides of the hydrocalumite type as efficient catalysts in various reactions. *Catalysis Today* 306:32–41.

Somani, P. R., S. P. Somani, and M. Umeno. 2006. Planer nano-graphenes from camphor by CVD. *Chemical Physics Letters* 430 (1–3):56–59.

Song, H., J. Liu, B. Liu, J. Wu, H.-M. Cheng, and F. Kang. 2018. Two-dimensional materials for thermal management applications. *Joule* 2 (3):442–463.

Song, X., J. Hu, and H. Zeng. 2013. Two-dimensional semiconductors: recent progress and future perspectives. *Journal of Materials Chemistry C* 1 (17):2952–2969.

Sun, P., Q. Chen, X. Li, H. Liu, K. Wang, M. Zhong, and T. Sasaki. 2015. Highly efficient quasi-static water desalination using monolayer graphene oxide/titania hybrid laminates. *NPG Asia Materials* 7 (2):e162.

Velický, M., and P. S. Toth. 2017. From two-dimensional materials to their heterostructures: an electrochemist's perspective. *Applied Materials Today* 8:68–103.

Wang, H., X. Xiang, and F. Li. 2010. Facile synthesis and novel electrocatalytic performance of nanostructured Ni–Al layered double hydroxide/carbon nanotube composites. *Journal of Materials Chemistry* 20 (19):3944–3952.

Wang, Q. H., K. Kalantar-Zadeh, A. Kis, J. N. Coleman, and M. S. Strano. 2012. Electronics and optoelectronics of two-dimensional transition metal dichalcogenides. *Nature Nanotechnology* 7 (11):699–712.

Wang, X., X. Yao, H. Bai, and Z. Zhang. 2022. Oxygen vacancy-rich 2D GO/BiOCl composite materials for enhanced photocatalytic performance and semiconductor energy band theory research. *Environmental Research* 212:113442.

Wu, L., D. Yang, G. Zhang, Z. Zhang, S. Zhang, A. Tang, and F. Pan. 2018. Fabrication and characterisation of Mg-M layered double hydroxide films on anodized magnesium alloy AZ31. *Applied Surface Science* 431:177–186.

Xiao, X., H. Song, S. Lin, Y. Zhou, X. Zhan, Z. Hu, and T. Li. 2016. Scalable salt-templated synthesis of two-dimensional transition metal oxides. *Nature Communications* 7 (1):1–8.

Xu, Y., W. Zhang, G. Zhou, M. Jin, and X. Li. 2022. In-situ growth of metal phosphide-black phosphorus heterojunction for highly selective and efficient photocatalytic carbon dioxide conversion. *Journal of Colloid and Interface Science* 616:641–648.

Yan, J., C. E. Ren, K. Maleski, C. B. Hatter, B. Anasori, P. Urbankowski, and Y. Gogotsi. 2017. Flexible MXene/graphene films for ultrafast supercapacitors with outstanding volumetric capacitance. *Advanced Functional Materials* 27 (30):1701264.

Yang, Y., Y. Peng, M. F. Saleem, Z. Chen, and W. Sun. 2022. Hexagonal boron nitride on III–V compounds: a review of the synthesis and applications. *Materials* 15 (13):4396.

Yu, H., H. Huang, K. Xu, W. Hao, Y. Guo, S. Wang, and Y. Zhang. 2017. Liquid-phase exfoliation into monolayered BiOBr nanosheets for photocatalytic oxidation and reduction. *ACS Sustainable Chemistry & Engineering* 5 (11):10499–10508.

Zavabeti, A., J. Z. Ou, B. J. Carey, N. Syed, R. Orrell-Trigg, E. L. Mayes, and R. B. Kaner. 2017. A liquid metal reaction environment for the room-temperature synthesis of atomically thin metal oxides. *Science* 358 (6361):332–335.

Zhang, Q., X. Li, Z. He, M. Xu, C. Jin, and X. Zhou. 2019. 2D semiconductors towards high-performance ultraviolet photodetection. *Journal of Physics D: Applied Physics* 52 (30):303002.

Zhou, K. G., N. N. Mao, H. X. Wang, Y. Peng, and H. L. Zhang. 2011. A mixed-solvent strategy for efficient exfoliation of inorganic graphene analogues. *Angewandte Chemie* 123 (46):11031–11034.

Zhu, B., B. Cheng, J. Fan, W. Ho, and J. Yu. 2021. g-C3N4-based 2D/2D composite heterojunction photocatalyst. *Small Structures* 2 (12):2100086.

Zhu, C., D. Du, and Y. Lin. 2017. Graphene-like 2D nanomaterial-based biointerfaces for biosensing applications. *Biosensors and Bioelectronics* 89:43–55.

3 Advances in Photocatalytic H$_2$ Generation using 2D Semiconductors

Irthasa Aazem, P. J. Jandas and Suresh C. Pillai

3.1 INTRODUCTION

Finding energy resources and value-added chemical feedstocks from eco-friendly renewable resource-based methods has particularly attracted researchers' attention today. The present scenario of extreme environmental pollution due to human intervention in the ecosystems as a part of unplanned urbanization and eco-hostile technology developments creates a massive imbalance between the CO_2 emission to the environment and its conversion through photosynthesis (Lokesh and Srivastava 2022). The accumulated CO_2 is considered as a primary reason for global warming and corresponding climatic imbalances happening worldwide. According to the UN, cyclonic storms and connected flood catastrophes across the globe have recorded a steep increase of around 134% since 2000, which is the outcome of unscientific human intervention in the ecosystem. Recent massive disasters in the USA, Pakistan, and places around the shores of the Bay of Bengal need to be considered in line with the above statistics. As a result, anything which can contribute to reducing the negative impact of global pollution is considered with utmost importance today. Therefore, technologies that can reduce the carbon footprint effect on the environment and methods/materials that proceed through green pathways are in high demand among the industrialists and research community (Ma et al. 2022).

Tremendous efforts are being made by every sector, from energy production/ utilization and industrial production, to reduce the harmful impact on the environment. Renewable resource-boned energy productions such as solar, wind, and wave energy-based electricity production are already finding their niche in the industry as substitutes for fossil fuel/or other non-environmentally friendly electricity production methods. However, the margin between the requirement and the production of clean energy has a huge gap and is increasing abruptly due to the fast urbanization/ increasing living dards across the world (Jiao et al. 2022; Nishiyama et al. 2021; Rao Pala and Peela 2021; Wu, Zhu, et al. 2022).

One of the most essential parts of fuel cell-based energy production is the generation and storage of H$_2$ via green methods (Jin et al. 2021; Chi et al. 2022; Yu

DOI: 10.1201/9781003343899-3

et al. 2015). A few methods are reported as suitable ways to produce H$_2$ from water and sunlight via photocatalytic methods. Given the abundance of both resources, hydrogen production by utilizing water and sunlight is considered environmentally friendly, efficient, and less complex. According to Lewis et al., the earth absorbs around 1022 J energy per day, which is way higher than the global demand for energy per annum at present (Lewis and Nocera 2006). The method of photocatalysis is greatly influenced by the natural photosynthesis process that happens in green vegetation and in a few microorganisms that convert energy from solar through the division of water. The history of photo-initiated water splitting to generate H$_2$ started with Fujishima and Honda's invention in 1972 (Fujishima and Honda 1972). They successfully did the photocatalytic water splitting using a TiO$_2$ catalyst. Since then, the area of research has grabbed considerable attention from energy and material scientists to find a scalable photocatalytic device for renewable energy production.

Since then, many research reports have been published on harvesting and transforming solar energy into suitable forms, including electrical and chemical energy, fuels, etc. (Wang, Vogel, et al. 2019; Murali et al. 2022; Kong et al. 2018). Various strategies like photovoltaic cells, which convert solar energy to electrical energy, become popular and commercialized the method even for big projects like an airport that is completely working through solar power. Today, researchers reached an energy conversion efficiency of more than 20% for solar cells. Still, the lack of advent in large-scale energy storage is a significant drawback in this field (Wang et al. 2022). Another aspect of utilizing solar energy is converting it into renewable fuels, considered green sources of energy (Tahir et al. 2022). Such solar fuels can be stored suitably and used per the requirement, making them practically more viable than direct solar to chemical energy converting devices. This can be possible mainly through two notable methods, CO$_2$ reduction into hydrocarbons and H$_2$ production (Sherryna and Tahir 2021).

H$_2$-based fuel cells are one promising category as it is termed clean and sustainable source of energy since they can yield a huge amount of power through ignition, 122 kJ mol^{-1}. This is way higher than that from any form of fossil fuel (Cao and Yu 2016). From this perspective, H$_2$ is identified as the most promising solution for clean harvesting energies and reducing anthropogenic carbon emissions to the environment. Hydrogen owns the utmost convertible energy level/unit mass by leaving water as the lone by-product. In this regard, H$_2$ is the cleanest, highest power generating, eco-friendly, and potential material to meet the impending requirement for energy (Kommula et al. 2022; Jiang et al. 2022; Wang, Li, and Domen 2019). However, the industrialization of new technology still must overcome many hurdles, including large-scale production, storage, and environmental aspects, including miniaturization, and regulating the hazardousness by avoiding the formation of any greenhouse gas. Reports are available based on the technologies like chemical, biological, thermo-chemical, electrolytic, photo electrolytic, and photocatalytic methods (Mian and Liu 2018; Jiang et al. 2018; Qian et al. 2018). Most of these methods rely on carbon feedstocks, and a large part of H$_2$ production today happens through natural gas via restructuring steam of CH$_4$ followed by the water gas shift method (Matos 2016). However, these methods could not be considered sustainable since one of the by-products is CO$_2$. Also, many of the methods mentioned above require high-pressure temperature conditions (Zhou, Jiang, et al. 2022). However, as stated above, the photo-initiated strategy utilizes only solar radiation, water, and potentially renewable raw materials. That makes the method competitive in comparison with conventional technologies.

3.2 PRINCIPLE OF PHOTOCATALYSIS

Photocatalytic water splitting on a semiconductor is one of the most researched areas in this part. The method involves three parts which are briefly depicted in Figure 3.1. In the initial step, light, energy/photons get absorbed by the semiconductor photocatalyst. Suppose the absorbed photons are associated with superior energy than the band gap energy of the photocatalytic semiconductor. In that case, the electrons (e^-) from the valence band (VB) get promoted to the conduction band (CB). This creates charged particles within the semiconductor by leaving +ve charged holes (h^+) in the VB when electrons get excited to the CB. In the second step, charge separation happens as the photogenerated pair of charges (e^-, h^+) move toward the exterior of the photocatalyst. Also, a few charge pairs recombine within the bulk of the semiconductor, known as volume recombination. In the final stage, surface redox reaction happens as photo-generated charges react with the reactants, again, some of the (e^-, h^+) pairs surface recombine in this step as well (Norouzi et al. 2019; Sutar, Otari, and Jadhav 2022; Che et al. 2018).

From this perspective, producing renewable energy by converting solar energy to chemical energy primarily depends on the efficiency of a semiconductor-based photocatalyst. A wide range of materials, including metal oxides, sulfides, nitrides, (oxy) sulfides, and (oxy) nitrides, with a metal ion of d^0 or d^{10} electronic configuration is being used for this application (Zhang et al. 2019). In addition to these materials, their hybrids, composites, covalent-organic frameworks, and metal-organic frameworks have been reported by many researchers (Chen et al. 2019). A reasonable light utilization rate is the prime requirement for a good photocatalyst for water splitting. Therefore, the band edges should be positioned suitably to meet the required potential for oxidation-reduction reaction. The wavelength region of the incident radiation within

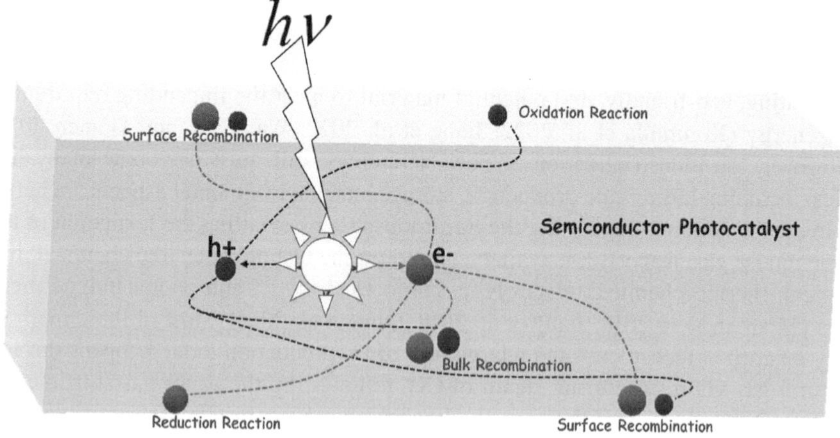

FIGURE 3.1 The basic principle behind semiconductor-based photocatalysis.

the visible-near-infrared region must be properly absorbed by the photocatalyst to get the electrons excited to the CB. Also, the photocatalyst must be efficient in separating light-generated charge carriers. Otherwise, it will result in poor photoactivity and quantum efficiency. The two popular methods to achieve such efficiency are developing a semiconductor system with a high light utilization rate and fabricating a combination (including co-catalyst) with great ability to separate photo-initiated e$^-$/h$^+$ pairs. The above-depicted new-age materials can act like an efficient photocatalyst of required properties of effective energy absorption from the solar spectrum and proficient charge separation capability through forming heterostructures (Zhou, Sun, et al. 2022).

However, few studies also suggest that the absorption of H$_2$O on the photocatalytic surface and its interaction with e$^-$/h$^+$ pairs play a crucial role in the outcome of a semiconductor rather than the above-discussed effects alone. However, such interaction depends on the grain surface, doping-controlled grain nucleation, porosity-affected H$_2$O adsorption/desorption kinetics, contact to the specific facets, and semiconductor/semiconductor or semiconductor/metal interfacial features. Also, a substantial size dependence and morphological sensitivity were reported in metal sulfide and metal oxide photocatalysts (Li et al. 2019) (Figure 3.2).

3.3 BASICS ABOUT PHOTOCATALYTIC HYDROGEN EVOLUTION

The H$_2$O dissociation was observed to be an endothermic reaction with possible products of H$_2$ and O$_2$, associated with a very high positive ΔG. The enthalpy of the reaction is estimated at 286 kJ mol^{-1}. Thermal dissociation of H$_2$O at a high temperature of around 2070 K can produce H$_2$ and O$_2$. However, using a semiconductor photocatalyst, it can allow the water to split at ambient temperatures. The primary mechanism behind the process was discussed previously in the earlier section. In

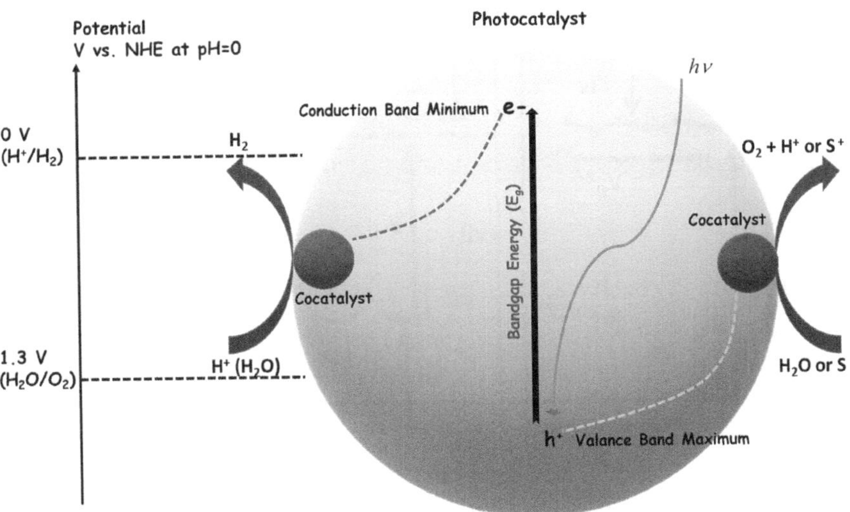

FIGURE 3.2 Schematic representation of energy diagram for H$_2$ production via dissociation of water on the surface of the photocatalytic semiconductor.

other words, a photocatalyst is a heterogeneous catalyst which can generate e^-/h^+ pairs by absorbing photons. These photogenerated charge pairs then contact the water molecules and co-catalysts on the photocatalyst surface for high energy ionic/free radical intermediates, followed by the formation of the final products H_2 by reduction of protons and O_2 by oxidation of water molecules at the electrodes (Wan et al. 2018).

Oxidation under the photocatalytic condition

$$H_2O + h^+_{vb} \text{-------} \longrightarrow 2H^+ + \tfrac{1}{2}\,O_2$$

Reduction under the photocatalytic condition

$$2H^+ + 2e^-_{CB} \text{-------} \longrightarrow H_2$$

Overall, water-splitting process

$$H_2O \xrightarrow{\quad TiO/h\nu \quad} H_2 + \tfrac{1}{2}\,O_2$$

The detailed energy transfer undergone during photocatalytic water splitting is given in Figure 3.3. The diagram is based on the common semiconducting materials associated with a wide bandgap like TiO_2 or $FeNbO_4$ (Mahvelati-Shamsabadi et al. 2021). The excitation of charges followed by radiative/non-radiative recombination of energetically excited particles can also occur during the process as an unwanted reaction.

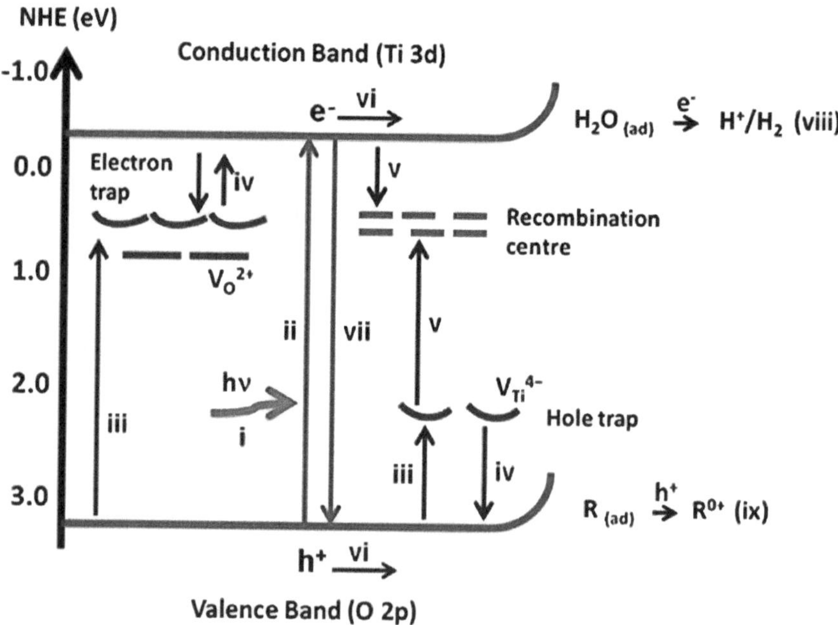

FIGURE 3.3 A scheme representing the likely energy transitions happening during the photoexcitation of TiO_2 (bandgap ~3.2 eV) (Gupta 2017).

This may happen in a minimal period of Femto-microseconds, which determines the quantum efficiency of the photocatalyst. The photon initiates may also interact with crystal imperfections or possible impurities in the lattice structure. In the following sections, the energy aspects of various stages are discussed.

3.3.1 Interacting Photons with Semiconductor Photocatalyst Surface: Importance of Morphology

When electromagnetic radiation is incident on a semiconductor surface, the energy can be reflected, absorbed, and transmitted based on the bandwidth of the catalyst, as given in Figure 3.4.

Considering the reflection phenomenon, it occurs when light scatters at the interphase of two media. The fraction of incident radiation reflected can be termed "R" and is determined as,

$R = I_r - I_0$, where I_0 is the intensity of the incident and I_r is the reflected radiation. The numerical R relies on the radiation's surface nature and incident angle (θ). On the other hand, absorption depends on the thickness of the semiconducting material. According to Bouguer's law,

$I = I_0.\exp(-\alpha.x)$, I - transmitted energy intensity, α - linear absorption coeffcient of semiconducting photocatalyst, x is the thickness.

The absorbed intensity of photons is expressed as

$$I_{abs} = (I - R)I_0.\exp(-\alpha.x)$$

From the equation, it is understandable that the incident radiation gets absorbed if the associated energy is identical or superior to the band gap energy, while if it is inferior, it gets reflected. In most of the reported cases of semiconductor-based photocatalysis, the nanomaterial-based catalyst gets exposed to a visible region of 200–800 nm

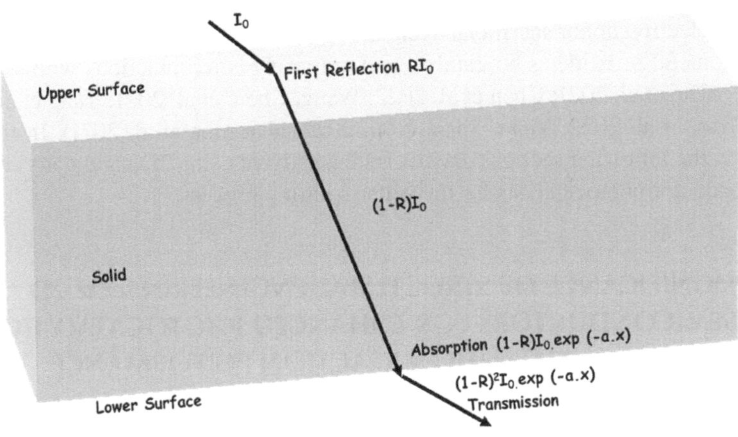

FIGURE 3.4 Reflection, absorption, and transmission of electromagnetic radiation on a semiconductor photocatalyst.

wavelength (1.5–6.5 eV). In such a situation, the processes of reflection and absorption are highly accountable. In this perspective, the size, surface porosity/roughness, refractive index, and reflectivity of the nanoparticle-based photoexcitation and the angle of incidence of the incident photon are crucial for the outcome of the photocatalysis process (Christoforidis and Fornasiero 2017).

3.3.2 Photoexcitation Process, the Entrapment, and Recombination of Charged Pairs (E^-/H^+)

The bulk region of the semiconductor is responsible for the second stage of the photo-splitting process. The electrons absorb energy and move to the CB, leaving holes in the valance band. This, followed by the diffusion of holes from one lattice point to another, leads to rapid radiative or non-radiative recombination. Radiative emission results in fluorescence and photoluminescence, while non-radiative emission will generate phonons. This process is followed by the redox reaction to the targeted products at the respective ends. The charged particle may trap within lattice defects and significantly influences the polarity and annihilation phenomenon. The possible types of defects are lattice vacancy (dislocation of an ion from the lattice site, additional atom/ion on the interstitial space, Frenkel defect, and Schottky defect). The energy levels corresponding to these defects are much closer to the valance and CB edge (Christoforidis and Fornasiero 2017).

3.3.3 The Transfer of Electron/Hole Pain on the Photocatalytic Surface

In this stage, the semiconductor-to-semiconductor and semiconductor-to-metal charge transfer effects play a crucial part as the interfacial transport of electrons within the CB and hole in the valance band results in photoreduction and photooxidation. However, this charge transfer highly relies on the energy variance between a photocatalyst's CB potential and the acceptor's reduction potential. But most of the time, rapid recombination and lattice entrapment due to defects affect the quantum efficiency significantly. In such cases, multiple strategies are used, including incorporation or electron donor sacrificial agent during the photoreduction, using a thin film of noble metal or oxide as co-catalyst and preparing heterojunctions with semiconductors (Cho et al. 2021; Chen et al. 2022; Wang, Chen, et al. 2021; Yang et al. 2016; Wu, Miao, et al. 2022; Yuan, Shen, et al. 2022; Hammon et al. 2021). In this perspective, the following sections discuss such an advent using 2D nanomaterial-based semiconductor photocatalysts for the water-splitting process.

3.4 SIGNIFICANCE OF STRUCTURAL ENGINEERING OF 2D SEMICONDUCTORS FOR ENHANCED PHOTOCATALYTIC HYDROGEN EVOLUTION REACTION PERFORMANCE

Structural and phase engineering of the 2D semiconductors is highly significant for utilizing their maximum potential for photocatalytic H_2 generation. Inertness to photo-corrosion and stability after a prolonged application is also highly crucial, along with optimal band structure. The most significant factor to be considered is the rate at which the electrons and holes re-join. The major reasons the recombination happens at a higher rate in photocatalysts are the higher penetration of the photons and short mean free paths of charge carriers. Among the various photocatalytic materials, 2D semiconductors exhibit certain merits such as sheet-like and ultrathin structure, adjustable number of layers, and high specific surface area (Saleem et al. 2020). By fine-tuning 2D semiconductors concerning the above-mentioned characteristics, their light harvesting, and catalytic activity can be enhanced.

The tunable number of layers favors the adjustment of the band gap and photo absorption rate. Also, by tuning the ultrathin sheet-like structures of these materials, enhancement of movement of charge carriers to the material's surface is made possible with a reduction in electron-hole pair recombination (Faraji et al. 2019). In addition to that, the enhancement of active sites at the surface is facilitated by the high surface area of these materials. An efficient photocatalyst should be able to absorb light even at low flux density. In the case of 2D semiconductors, the tunability of their layered and thin structure enables this and thus the charge carrier transport distance is shortened. Moreover, by achieving the retainment of electrons in the ultrathin layer, considerable enhancement in optical and electronic properties is obtained. Another main factor contributing to the possibility of engineering 2D semiconductors is the availability of strong in-plane covalent bonds. The presence of covalent bonds sets a platform for the formation of semiconductor structures with the chemical composition varying with position Various approaches have been experimented with by tuning the chemical arrangement, molecular orientation, and morphology of the 2D semiconductors to achieve their maximum photocatalytic hydrogen evolution reaction (HER) efficiency. Defect engineering and modulating the electronic structure of the materials, surface engineering, and surface patterning methods, compositing or doping with a different class of materials and morphology controllable synthesis techniques have been extensively studied (Ganguly et al. 2019).

The major contribution of defect engineering to photocatalytic HER is favoring the relocation of photogenerated charge carriers by generating electron tap sites and thereby reducing the materials band gap (Shu et al. 2017; Hu et al. 2017). Apart from that, additional energy levels could be introduced through defect engineering. 1T-MoS$_2$ nanosheets incorporated with a large number of oxygen defects (DRM) were synthesized and formed heterostructure with Cadmium disulfide (CdS) nanorods for photocatalytic H$_2$ production by Zhang, Li et al. (2018). The defect sites of this material were present abundantly in the edges and basal planes thereby improving the active site density. The DRM CdS heterostructure exhibited an HER of 132. 4 mmol h^{-1} g^{-1} at $\lambda = 420$ nm, which is 1.3 times larger than that of the defect-free DRM CdS heterostructure. Moreover, when compared with the H$_2$ production rate of bare CdS (20 mmol h^{-1} g^{-1}), the DRM CdS heterostructure shows superior activity. The availability of oxygen in MoS$_2$ improves conductivity, minimizes the energy barrier for H$_2$ evolution, and increases the catalytic stability of the material. Abundant Zn vacancies containing ZnS/g-C$_3$N$_4$ heterostructure were synthesized by Hao et al. (2018) in a study about the assisted enhancement of charge uncoupling in photocatalysts. The study showed H$_2$ evolution rate exceeding 30 times as compared to that of g-C$_3$N$_4$ under photoirradiation conditions with an increased stability. The improved activity was ascribed to the strong contact at the interface between ZnS and g-C$_3$N$_4$. In a different work on C$_3$N$_4$ by Zhang, Lu et al. (2018), pore-induced graphitic carbon nitride (Pg-C$_3$N$_4$) was prepared with defect sites for excellent HER activity. By utilizing a porous kaolinite clay template for synthesizing Pg-C$_3$N$_4$, edge site defects were obtained. Here, during the photocatalysis reaction, photo-excited holes were suppressed and photo electron reaction was enhanced. The PHE rate for Pg-C$_3$N$_4$ was observed to be 1917 µmol^{-1} g^{-1} h^{-1}, which was 2.37 folds as that of g-C$_3$N$_4$ under visible light irradiation.

Surface modification through surface engineering is another strategy for fine-tuning the physicochemical properties of nanomaterials by introducing structural defects. Various surface engineering strategies include the interaction of the surface with a co-catalyst (Zhou et al. 2021), interfacial engineering of 2D semiconductor nanostructures (Zhang, Qian, et al. 2018), active site enhancement through modified synthesis routes (Bie et al. 2018), modulating the external crystal facets (Yuan et al. 2017), etc. Excellent charge transfer performance, band gap improvement, enhancement of active site and surface carrier density, improved magnetic properties, etc. could be achieved through surface engineering. The co-catalyst-based surface modification of the photocatalyst material primarily modulates the reaction kinetics of surface redox reactions by decreasing the activation energy barrier (Zhang, Lan, and Wang 2017). In a study by Maeda et al. (2009), the photocatalytic water-splitting capability of graphitic carbon nitride (g-C$_3$N$_4$) was found to be enhanced when modified with a platinum (Pt) co-catalyst. The density of active sites for H$_2$ evolution was found to be enhanced with the increase in the content of Pt while the excess amount of Pt hindered the light absorption of the photocatalyst and resulted in increased electron-hole recombination. Apart from that, the good dispersion of Pt deposits also favored the enhancement of the activity of the photocatalyst. However, since upscaling is a big challenge with noble metal co-catalysts, inexpensive co-catalysts for HER are in higher demand. Mo–O$_2$–C nanowire clusters were introduced as a noble-metal-free co-catalyst for photocatalytic HER in a study by Chen et al. (2018). A 157.5 times improvement in property was observed with a hydrogen evolution rate of 1071.0 µmol g^{-1} h^{-1}. The improved activity was because of the increased density of surface-active sites and the effective transfer of photoelectrons. Surface modification of photocatalyst through interfacial engineering contributes to photo-induced charge extraction and regulates the surface charge states of the photocatalyst. In a study by Chava, Son, and Kang (2021), heterojunctions of Bi/Bi$_2$MoO$_6$-MoS$_2$ were grown over CdS nanorods by a solvothermal reaction to obtain a single nanoarchitecture of CdS-Bi/Bi$_2$MoO$_6$-MoS$_2$. Enhanced visible light harnessing and efficient decoupling of electron-hole pairs was achieved to show an H$_2$-evolution activity of 2185 µmol g^{-1} h^{-1} which is 10–15 times greater than that of the individual components of the heterostructure. Moreover, the charge separation was greatly enhanced via the z-scheme mechanism by the integrated metallic Bi between CdS and BiMoO.

Another primary strategy for structural engineering in 2D semiconductor photocatalysts is compositing with other materials. Embedding metal nanoparticles on the surface of semiconductors is an effective tailoring process for enhancing photocatalytic behavior. The overlay of the wavefunction of the semiconductor/metal interface occurs during excitation and causes bending in the semiconductor band structure. Also, an energy barrier is created to prevent the backflow of electrons to the semiconductor. This will largely contribute to preventing the recombination of charge carriers and promotes the enhancement of the photocatalytic H$_2$ evolution. Metal-modified carbon quantum dots (CDs) were combined with CdS as a co-catalyst for photocatalytic HER in a study by Wang, Chen et al. (2019). Among the various doped metals, Bi-doped CDs/CdS composite exhibited the best activity with optimum interfacial charge separation. The HER of Bi-doped CDs/CdS was 4.2 times that of the undoped CDs/CdS. In another work exploring the conversion of basal planes of transition metal dichalcogenides (TMDs) to catalytically active sites, Cu-doped MoS sheets were incorporated over CdS nanorods by Hong et al. (2017). The hydrogen production rate of the developed material was observed to be 52 times increased in comparison to the activity of pristine CdS photocatalyst. This was attributed to the surface shuttling characteristics and effective charge separation obtained through the combined effect of metal doping and MoS$_2$. Moreover, the occurrence of new active sites and edge sites was observed in the materials. One significant advantage of compositing different semiconducting materials is that it opens the possibility of forming heterojunctions, enhancing the photocatalytic HER. A study by Wei et al. (2021) demonstrated the formation of p-n heterojunction with excellent photocatalytic hydrogen evolution potential by compositing 2D NiO with 1D cadmium sulfide (CdS). The resultant composite contained CdS nanorods uniformly anchored onto 2D nanosheets of NiO (2D/1D NiO/CdS). The NiO/CdS compound exhibited a value of 1300 µmol g^{-1} h^{-1} for the photocatalytic hydrogen generation rate, which is eight-folds as that of the CdS nanorods. Efficient separation of photogenerated electrons was achieved by forming p-n NiO/CdS heterojunction. This further led to upscaling of the H$_2$ production efficiency under visible light, apart from that this material showed outstanding stability over four cycles of H$_2$ production. Liu et al. (2017), in a different work, prepared a composite of 0D colloidal quantum dots (CQDs) and 2D MoS$_2$ by which the catalytic performance of MoS$_2$ and light harvesting capability of CQDs combinedly contributed to the excellent H$_2$ production property. Also, efficient transport of charge carriers between ZAIS CQDs and MoS$_2$ was observed. A Zn-Ag-In-S (ZAIS) CQD was developed and formed a nanocomposite MoS$_2$ to achieve an external quantum efficiency of 40.8% at 400 nm. Moreover, a hydrogen generation rate of 40.1 µmol g^{-1} h^{-1} was observed under visible light irradiation conditions.

3.5 ADVANCED 2D SEMICONDUCTOR MATERIALS DEVELOPED FOR PHOTOCATALYTIC HER PERFORMANCE

3.5.1 GRAPHITIC CARBON NITRIDE (g-C$_3$N$_4$)

Graphitic carbon nitride is a polymeric semiconductor which exhibits excellent photocatalytic activity and a suitable band structure for HER. This material is visible-light-active with a narrow band gap of ca. 2.7 eV and an optical wavelength of ca. 460 nm (Wang et al. 2009). Engrossing electronic band structure, simple synthesis and functionalization routes, ease of compositing with a different class of materials, and excellent thermal and chemical stability makes g-C$_3$N$_4$ a promising material for next-generation photocatalytic HER. However, certain disadvantages such as high charge carrier recombination rates and low electrical conductivity and the absence of photocatalytic activity above 460 nm limit their practical applications to a great extent (Ye et al. 2015). To overcome these drawbacks, certain strategies like making them highly crystalline, doping, and imparting porosity have been investigated and the preparation of g-C_3N_4 composite heterojunctions has been implemented (Han et al. 2016; Xiong et al. 2016). Since 2009, when g-C_3N_4 was first introduced as a photocatalyst for HER by Wang et al. (2009), the synthesis of g-C_3N_4 has been carried out mainly through the pyrolysis of precursors rich in nitrogen (Zhou et al. 2015; Bai et al. 2013). Figure 3.5 shows the schematic of different synthesis routes of g-C_3N_4 from nitrogen-rich precursors and their photocatalytic water-splitting performances. Song and co-workers devised g-C_3N_4 by simple calcination of urea and were composited with polyoxoniobate to form a binary type-II heterojunction for photocatalytic HER application (Song et al. 2022). In another work, Xiao et al. synthesized tubular g-C_3N_4 from melamine precursor and phosphoric acid through a hydrothermal reaction (Xiao et al. 2022). Various strategies used for the synthesis of g-C_3N_4 from N-rich compounds include microwave-assisted heating methods (Yuan et al. 2014), template-assisted technique (Yan 2012), molten state synthesis (Bojdys et al. 2008), and ionic liquid techniques (Xu et al. 2018). As for the electronic structure of g-C_3N_4, the lone pair electrons of N are in charge of the formation of the VB and its band structure (Liao et al. 2019). Moreover, the N P_0 and C P_z orbitals are responsible for the valence and CBs and the oxidation and reduction for water splitting happen in the N atom and C atom (Lotsch et al. 2007).

In a recent study, praus et al. synthesized g-C_3N_4 with two band gaps of 2.14 and 2.47 eV from melamine in the argon atmosphere and found to have three times the photocatalytic HER performance (1547 moil g^{-1}) as that of the g-C_3N_4 synthesized in the oxygen atmosphere (547 μmol g^{-1}). The lower band gap formed because of the vacancies formed was attributed to the argon atmosphere and this was beneficial for H_2 evolution (Praus et al. 2022). Lu et al. demonstrated the positive effect of adding carbonaceous materials on the photocatalytic activity of g-C_3N_4. In this work, an eco-friendly bamboo-charcoal-loaded g-C_3N_4 was developed, which enhanced the electron-hole pair separation. An increased HE activity of 4.1 mmol $g^{-1}h^{-1}$ which is 2.3 folds as that of bare g-C_3N_4 was observed along with an increment in surface area from 85 to 120 m^2g^{-1}. Moreover, the presence of an ohmic interaction between g-C_3N_4 and bamboo-charcoal facilitated the fast movement of photogenerated charge carriers with reduced recombination (Lu et al. 2022). Wang et al. developed a g-C_3N_4 with enhanced electron active sites on the surface by simple two-step calcination of melamine. The accumulated electron sites (BN_2) present on the surface of g-C_3N_4 enhanced its photocatalytic hydrogen evolution activity by 2.3 times under 420 nm, as that of the pristine g-$C_3N_4.$ The decrease in interlamellar distance and potential energy barrier of the interlayers favored the quick transport of charge in the z-axis. This in turn improved water adsorption. Additionally, the recombination rate of the electron-hole pair was reduced by the enhancement of π–π conjugation. In another work by Yuan, Fang et al. (2022) for the enhancement of water-splitting reaction, the photocatalytic property of g-C_3N_4 was boosted by compositing it with $MoS_2.$ Here, the typical laminated structure of a few-layered MoS_2 was highly favorable for the separation of photogenerated electron-hole pair. Hexagonal tubes of g-C_3N_4

were also coupled with Co$_3$S$_4$ to form heterojunctions to boost the spatial charge separation (Wang, Hao, et al. 2021). This enhanced the number of electrons with higher reduction ability and enhanced the H$_2$ evolution reaction. The ultrafine Co$_3$S$_4$ nanoparticles enhanced the harnessing of light through increased scattering and reflection and improved the charge separation. Compared to the pristine g-C$_3$N$_4$, the activity of Co$_3$S$_4$ modified g-C$_3$N$_4$ was increased 176 times (2120 µmol g^{-1} h^{-1}).

3.5.2 TMDs

TMDs are a class of lamellar materials that fall under the general formula MX$_2$ where M denotes a transition metal and of group 4-10 and X represents a chalcogen atom. Certain TMDs such as WS$_2$ and MoS$_2$ nanosheets exhibit thickness and lateral size-dependent tunable energy band gaps suitable for solar energy harvesting (Wilson and Yoffe 1969; Wang et al. 2012). Additionally, these materials with narrow band gaps help in extending the absorption of other semiconductors in the visible region (Tang et al. 2017). A van der Waals heterostructure of C$_2$N/WS$_2$ was synthesized by Kumar et al. for water splitting (Kumar, Das, and Singh 2018). Water oxidation and water reduction were obtained on the WS$_2$ and C$_2$N layers, respectively, upon exposure to light. The separation of photoactivated electrons was obtained in the type – II heterostructure formed by the transport of charges from WS$_2$ to C$_2$N monolayer which also served the purpose of extending the lifetime. TMDs have been proven to be suitable replacements for classic co-catalyst materials such as Pt which are expensive and less abundant for materials in water splitting. The excellent co-catalytic activity of TMDs has been demonstrated in recent studies. WS$_2$ nanoparticles covered with MoSe$_2$ nanosheets were synthesized using a simple hydrothermal synthesis method by Padma et al. In comparison to the pristine WS$_2$ nanoparticles, the WS$_2$/MoSe$_2$ composites exhibited higher photocatalytic activity. Under white light LED irradiation, the composite showed an activity of 1600.2 µmol g^{-1}h^{-1} which is ten times that of the pristine WS$_2$.

Another factor which makes TMDs suitable for water splitting is their ability for charge separation and to act as charge transporters. These two properties co-exist in TMDs. TMDs contain 2H and 1T crystal phases which are responsible for light harvesting and charge transport respectively in these materials. Mahler et al. (2014) demonstrated the preparation through a colloidal synthesis technique for photocatalytic HER. Controlled growth of the crystal phase was presented which resulted in obtaining the distorted metallic 1T-WS$_2$ along with the common semiconducting structure of 2H-WS$_2$. This work also showed a three-fold increase in the photocatalytic activity of TiO$_2$ when it was composited with a distorted 1T-WS$_2$ structure. In photocatalytic reactions, the lifetime of photogenerated charge carriers is a crucial factor. In the case of 2H WS$_2$ and MoS$_2$, the electron-hole pair recombination rate is very high. Developing nanosheets of these materials with suitable architectures and compositing them with other semiconducting materials helps to increase their lifetime. In a study by Raja et al., metallic-type MoS$_2$ (1T) was grown over TiO$_2$-Cds. The passivation of 1T-MoS$_2$ sheets at the interface helped in reducing the charge recombination through uncovered CdS sites. The shorter band gap and increasing electron transport framework resulted in increasing the photocatalytic activity of the material (Figure 3.6).

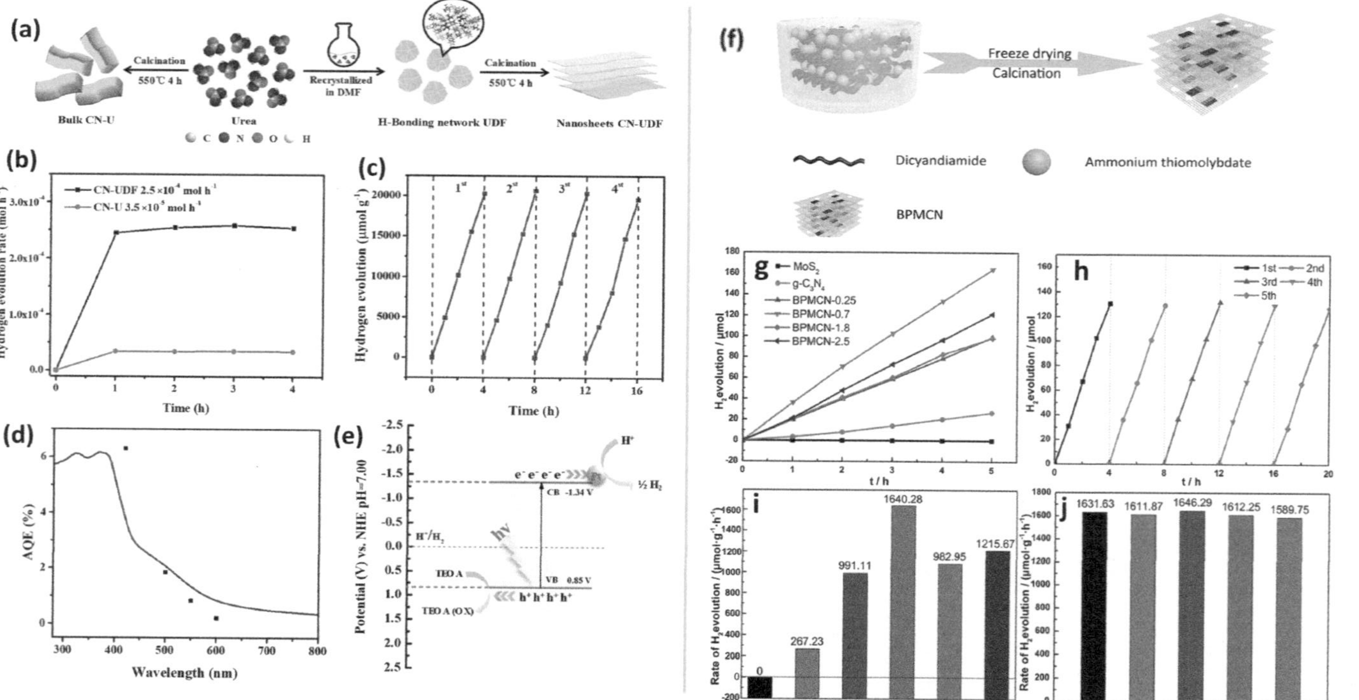

FIGURE 3.5 Various synthesis approaches of g-C$_3$N$_4$ from nitrogen-rich precursors and their photocatalytic water splitting performances. (a) schematic illustration synthesis of g-C$_3$N$_4$ nanosheets from a recrystallized urea-based precursor (CN-UDF), (b) Photocatalytic hydrogen splitting activity of CN-UDF), (c) prolonged stability of CN-UDF), (d) wave-length dependent AQEs of CN-UDF), (e) Proposed mechanism of activity of CN-UDF. Reproduced with permission from copyright (Zhang et al. 2020). (f) Schematic representation of the synthesis of a porous MoS$_2$/g-C$_3$N$_4$ (BPMCN) heterojunction photocatalysts, (g) and (i) hydrogen evolution rates of Photocatalytic activity of MoS$_2$ and BPMCN, (h) and (j) recycling test of BMPCN. Reproduced with permission from (Cao et al. 2017).

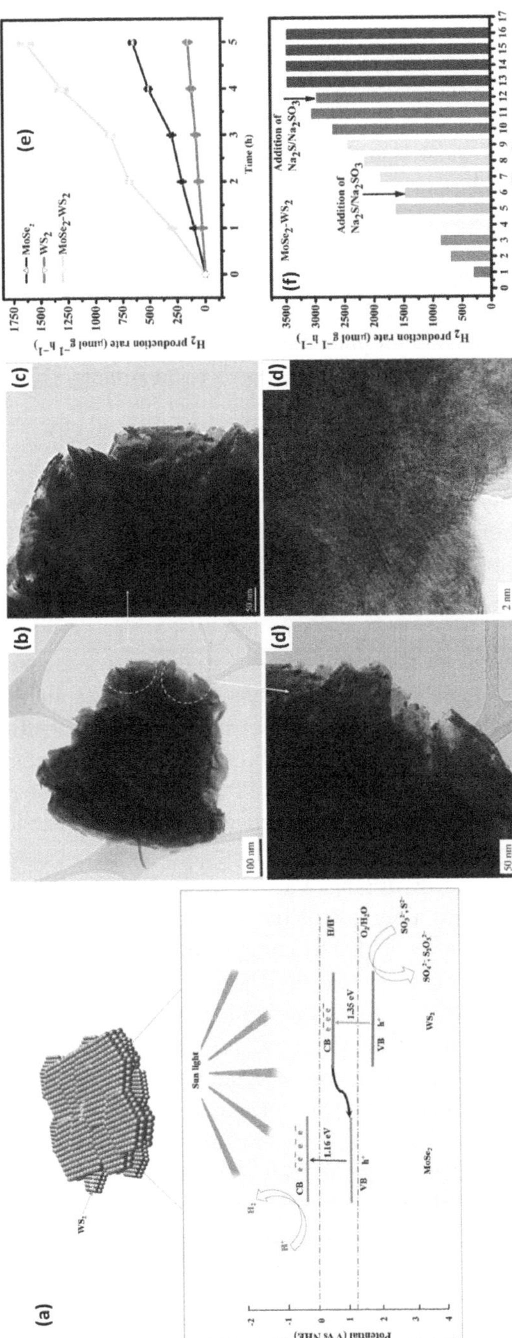

FIGURE 3.6 Photocatalytic hydrogen production mechanism and performance of MoSe$_2$-WS$_2$ heterojunction composite: (a) Scheme of the mechanism of action of MoSe$_2$-WS$_2$ nanostructure photo irradiation. (b–d) TEN images of MoSe$_2$-WS$_2$ nanocomposite. (e) HRTEM image of MoSe$_2$-WS$_2$. (f) Hydrogen production rate of MoSe$_2$, WS$_2$ and MoSe$_2$-WS$_2$ nanocomposite. (g) Visible light activated cycling stability of the composite catalyst. Reproduced with permission from (Padma et al. 2022).

By facilitating a strong interfacial bond between TMDs and the compositing materials a strong electronic interaction can be formed at the interfaces which significantly enhance the charge separator and charge transporter role of TMDs. This is mostly contributed by the nature of heterostructures, and heterojunctions formed by TMDs with the combining materials. The impact of MoS_2 on plasmon-enhanced photocatalytic HER was investigated on Au nanoparticles @ MoS_2 core-shell structure by Yang, Guo, and Li (2017). By tuning the MOS_2 coverage time on the Au nanoparticles by changing the hydrothermal reaction time from 3–18 h, it was found that a moderate reaction time enables the penetration of water molecules through MOS_2 to reach the interface of the core-shell. This also led to an excellent Hydrogen generation of 2110.7 µmol. Moreover, the yield of hydrogen gas was found to be higher than that of pristine MoS_2 nanospheres (880.0 µmol g^{-1}). Charge separation at the interface is another important requirement for photocatalytic water splitting; the incorporation of materials that impart high photostability is also a good strategy (Figure 3.7).

3.6 CONCLUSIONS AND PERSPECTIVES

As far as the photocatalytic HER is concerned, 2D semiconductor materials and their composites are crucial owing to their highly favorable properties such as large number of active sites, appropriate band energy levels, and charge transport properties for this application. 2D semiconductor materials such as polymeric graphitic carbon nitride (g-C_3N_4), TMDs, and graphene have been proven to be an excellent choice of materials because of their appropriate band gap, good charge transport capability, high density of active sites, good physicochemical stability, and suitable band energy levels. However, they also face many setbacks in the practical scenario. Particularly, the inherent thermodynamic instability suffered by 2D semiconductors is a serious drawback and therefore the long-term stability of these materials needs to be addressed for making them suitable for practical applications. The low light harnessing capability of these materials owing to their thin layered structure is a huge concern. Therefore, coupling or doping with other semiconductor photocatalytic materials should be carried out. Poor conductivity of nanosheets is a huge drawback in the case of TMDs. This can be tackled by compositing these materials with highly conducting materials like graphene and engineering the external phase of these materials to impart active sites. Moreover, the formation of heterojunctions with suitable metal oxide semiconductors is also a successful strategy. More importantly, the controlled synthesis of TMDs for achieving selected morphologies and dimensional stabilities is also extremely crucial as the presence of active sites and edges is the key factor responsible for the photocatalytic activity of these materials. Diverse experimental and modeling studies investigating new 2D semiconducting materials such as TaS_2, TiS_2, WSe_2, and $MoSe_2$ have been reported. Single-layer monochalcogenides such as GaSe, InS, etc. and single-layer metal-phosphorus-trichalcogenides have exhibited electronic properties suitable for photo-electrocatalytic water splitting for hydrogen production. However, efficient and economic fabrication strategies for the mass production of these materials are to be explored.

Even though the electronic structure and defect engineering strategies in 2D semiconductors and their composites with different classes of materials have helped a

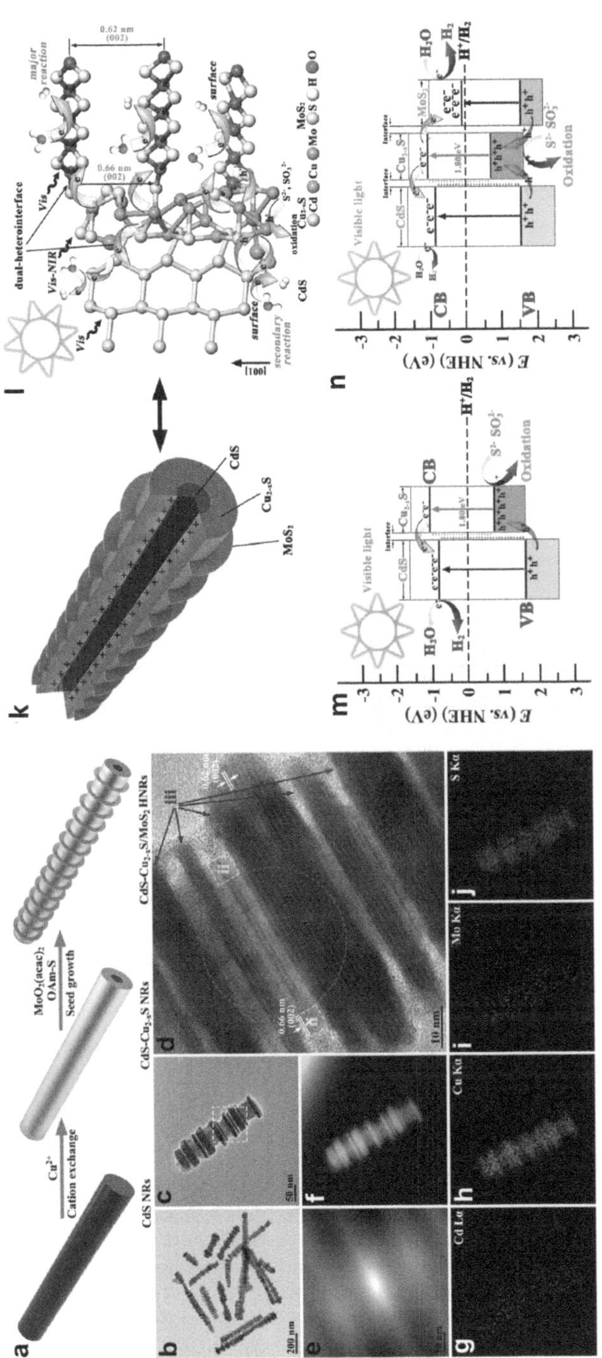

FIGURE 3.7 Structural and elemental characterization and proposed photocatalytic activity mechanism of CdS-Cu$_2$–xS/MoS$_2$ composite. (a) Schematic of formation procedure of the composite, (b–f) Morphological analysis of the materials, (g–j) Elemental mapping of the material. (k–n) Schematic of the crystal structure of the heterojunction. Reproduced with permission from (Liu et al. 2020).

lot to improve their photocatalytic HER, the underlying mechanisms and complete potential of these materials are yet to be fully explored. The limitations associated with existing experimental protocols project the need for reliable theoretical calculations and simulation techniques to understand and predict the photocatalytic efficiency of these materials. Apart from that, new material design strategies and performance evaluation methods suitable for investigations afar laboratory scale are to be implemented. Additionally, economic scale-up of the existing synthesis techniques for mass production of the photocatalytic materials to meet the requirements in the practical scenario through eco-friendly methods are to be discovered.

REFERENCES

Bai, Xiaojuan, Li Wang, Ruilong Zong, and Yongfa Zhu. 2013. Photocatalytic activity enhanced via g-C$_3$N$_4$ nanoplates to nanorods. *The Journal of Physical Chemistry C* 117 (19):9952–9961.

Bie, Chuanbiao, Junwei Fu, Bei Cheng, and Liuyang Zhang. 2018. Ultrathin CdS nanosheets with tunable thickness and efficient photocatalytic hydrogen generation. *Applied Surface Science* 462:606–614.

Bojdys, Michael J., Jens-Oliver Müller, Markus Antonietti, and Arne Thomas. 2008. Ionothermal synthesis of crystalline, condensed, graphitic carbon nitride. *Chemistry* 14 (27):8177–8182.

Cao, Shaowen, and Jiaguo Yu. 2016. Carbon-based H$_2$-production photocatalytic materials. *Journal of Photochemistry and Photobiology C: Photochemistry Reviews* 27:72–99.

Cao, Youzhi, Qin Gao, Qiao Li, Xinbo Jing, Shufen Wang, and Wei Wang. 2017. Synthesis of 3D porous MoS$_2$/g-C$_3$N$_4$ heterojunction as a high efficiency photocatalyst for boosting H$_2$ evolution activity. *RSC Advances* 7 (65):40727–40733.

Chava, Rama Krishna, Namgyu Son, and Misook Kang. 2021. Surface engineering of CdS with ternary Bi/Bi$_2$MoO$_6$-MoS$_2$ heterojunctions for enhanced photoexcited charge separation in solar-driven hydrogen evolution reaction. *Applied Surface Science* 565:150601.

Che, Yuping, Bingxin Lu, Qi Qi, et al. 2018. Bio-inspired Z-scheme g-C$_3$N$_4$/Ag$_2$CrO$_4$ for efficient visible-light photocatalytic hydrogen generation. *Scientific Reports* 8 (1):16504.

Chen, Dan, Xiaonan Wang, Xiaoqiao Zhang, Yan Yang, Yao Xu, and Guangren Qian. 2019. Facile fabrication of mesoporous biochar/ZnFe$_2$O$_4$ composite with enhanced visible-light photocatalytic hydrogen evolution. *International Journal of Hydrogen Energy* 44 (36):19967–19977.

Chen, Jun, Si-Jia Wu, Wen-Jun Cui, et al. 2022. Nickel clusters accelerating hierarchical zinc indium sulfide nanoflowers for unprecedented visible-light hydrogen production. *Journal of Colloid and Interface Science* 608:504–512.

Chen, Zhigang, Kaixiang Xia, Xiaojie She, et al. 2018. 1D metallic MoO$_2$-C as co-catalyst on 2D g-C$_3$N$_4$ semiconductor to promote photocatlaytic hydrogen production. *Applied Surface Science* 447:732–739.

Chi, Manzhou, Huijie Li, Xing Xin, Lin Qin, Hongjin Lv, and Guo-Yu Yang. 2022. All-inorganic bis-Sb$_3$O$_3$-functionalized A-type Anderson–Evans polyoxometalate for visible-light-driven hydrogen production. *Inorganic Chemistry* 61 (22):8467–8476.

Cho, Hyekyung, Hyunku Joo, Hansung Kim, Ji-Eun Kim, Kyoung-Soo Kang, and Jaekyung Yoon. 2021. Improved photoelectrochemical properties of TiO$_2$ nanotubes doped with Er and effects on hydrogen production from water splitting. *Chemosphere* 267:129289.

Christoforidis, Konstantinos C., and Paolo Fornasiero. 2017. Photocatalytic hydrogen production: a rift into the future energy supply. *ChemCatChem* 9 (9):1523–1544.

Faraji, Monireh, Mahdieh Yousefi, Samira Yousefzadeh, et al. 2019. Two-dimensional materials in semiconductor photoelectrocatalytic systems for water splitting. *Energy & Environmental Science* 12 (1):59–95.

Fujishima, Akira, and Kenichi Honda. 1972. Electrochemical photolysis of water at a semiconductor electrode. *Nature* 238 (5358):37–38.

Ganguly, Priyanka, Moussab Harb, Zhen Cao, et al. 2019. 2D nanomaterials for photocatalytic hydrogen production. *ACS Energy Letters* 4 (7):1687–1709.

Gupta, Narendra M. 2017. Factors affecting the efficiency of a water splitting photocatalyst: a perspective. *Renewable and Sustainable Energy Reviews* 71:585–601.

Hammon, Sebastian, Mara Klarner, Gerald Hörner, et al. 2021. Combining metal nanoparticles with an Ir(III) photosensitizer. *The Journal of Physical Chemistry C* 125 (46):25765–25773.

Han, Qing, Bing Wang, Jian Gao, et al. 2016. Atomically thin mesoporous nanomesh of graphitic C$_3$N$_4$ for high-efficiency photocatalytic hydrogen evolution. *ACS Nano* 10 (2):2745–2751.

Hao, Xuqiang, Jun Zhou, Zhiwei Cui, Yicong Wang, Ying Wang, and Zhigang Zou. 2018. Zn-vacancy mediated electron-hole separation in ZnS/g-C$_3$N$_4$ heterojunction for efficient visible-light photocatalytic hydrogen production. *Applied Catalysis B: Environmental* 229:41–51.

Hong, Sangyeob, D. Praveen Kumar, Eun Hwa Kim, et al. 2017. Earth abundant transition metal-doped few-layered MoS$_2$ nanosheets on CdS nanorods for ultra-efficient photocatalytic hydrogen production. *Journal of Materials Chemistry A* 5 (39):20851–20859.

Hu, Wei, Lin Lin, Ruiqi Zhang, Chao Yang, and Jinlong Yang. 2017. Highly efficient photocatalytic water splitting over edge-modified phosphorene nanoribbons. *Journal of the American Chemical Society* 139 (43):15429–15436.

Jiang, Xia, Yan-Xin Chen, Jing-Wen Zhou, Shi-Wei Lin, and Can-Zhong Lu. 2022. Pollen carbon-based rare-earth composite material for highly efficient photocatalytic hydrogen production from ethanol–water mixtures. *ACS Omega* 7 (34):30495–30503.

Jiang, Zhifeng, Hongli Sun, Tianqi Wang, et al. 2018. Nature-based catalyst for visible-light-driven photocatalytic CO2 reduction. *Energy & Environmental Science* 11 (9):2382–2389.

Jiao, Haimiao, Chao Wang, Lunqiao Xiong, and Junwang Tang. 2022. Insights on carbon neutrality by photocatalytic conversion of small molecules into value-added chemicals or fuels. *Accounts of Materials Research* 3 (12):1206–1219.

Jin, Zhiliang, Xiangyi Wang, Yuanpeng Wang, Teng Yan, and Xuqiang Hao. 2021. Snowflake-like Cu$_2$S coated with NiAl-LDH forms a p–n heterojunction for efficient photocatalytic hydrogen evolution. *ACS Applied Energy Materials* 4 (12):14220–14231.

Kommula, Bramhaiah, Pandiselvi Durairaj, Samita Mishra, et al. 2022. Self-assembled oligothiophenes for photocatalytic hydrogen production and simultaneous organic transformation. *ACS Applied Nano Materials* 5 (10):14746–14758.

Kong, Dan, Yun Zheng, Marcin Kobielusz, et al. 2018. Recent advances in visible light-driven water oxidation and reduction in suspension systems. *Materials Today* 21 (8):897–924.

Kumar, Ritesh, Deya Das, and Abhishek Kumar Singh. 2018. C$_2$N/WS$_2$ van der Waals type-II heterostructure as a promising water splitting photocatalyst. *Journal of Catalysis* 359:143–150.

Lewis, Nathan S., and Daniel G. Nocera. 2006. Powering the planet: chemical challenges in solar energy utilisation. *Proceedings of the National Academy of Sciences of the United States of America* 103 (43):15729–15735.

Li, Lingjiao, Hai Yu, Jing Xu, Sheng Zhao, Zeying Liu, and Yanru Li. 2019. Rare earth element, Sm, modified graphite phase carbon nitride heterostructure for photocatalytic hydrogen production. *New Journal of Chemistry* 43 (4):1716–1724.

Liao, Guangfu, Yan Gong, Li Zhang, Haiyang Gao, Guan-Jun Yang, and Baizeng Fang. 2019. Semiconductor polymeric graphitic carbon nitride photocatalysts: the "holy grail" for the photocatalytic hydrogen evolution reaction under visible light. *Energy & Environmental Science* 12 (7):2080–2147.

Liu, Guoning, Charles Kolodziej, Rong Jin, et al. 2020. MoS$_2$-stratified CdS-Cu$_2$–xS core–shell nanorods for highly efficient photocatalytic hydrogen production. *ACS Nano* 14 (5):5468–5479.

Liu, Xiao-Yuan, Hao Chen, Ruili Wang, et al. 2017. 0D–2D quantum dot: metal dichalcogenide nanocomposite photocatalyst achieves efficient hydrogen generation. *Advanced Materials* 29 (22):1605646.

Lokesh, Sankhula, and Rohit Srivastava. 2022. Advanced two-dimensional materials for green hydrogen generation: strategies toward corrosion resistance seawater electrolysis-review and future perspectives. *Energy & Fuels* 36 (22):13417–13450.

Lotsch, Bettina V., Markus Döblinger, Jan Sehnert, et al. 2007. Unmasking melon by a complementary approach employing electron diffraction, solid-state NMR spectroscopy, and theoretical calculations—structural characterisation of a carbon nitride polymer. 13 (17):4969–4980.

Lu, Yongfeng, Wensong Wang, Hongrui Cheng, et al. 2022. Bamboo-charcoal-loaded graphitic carbon nitride for photocatalytic hydrogen evolution. *International Journal of Hydrogen Energy* 47 (6):3733–3740.

Ma, Zili, Yihong Chen, Chao Gao, and Yujie Xiong. 2022. A minireview on the role of cocatalysts in semiconductor-based photocatalytic CH$_4$ conversion. *Energy & Fuels* 36 (19):11428–11442.

Maeda, Kazuhiko, Xinchen Wang, Yasushi Nishihara, Daling Lu, Markus Antonietti, and Kazunari Domen. 2009. Photocatalytic activities of graphitic carbon nitride powder for water reduction and oxidation under visible light. *The Journal of Physical Chemistry C* 113 (12):4940–4947.

Mahler, Benoit, Veronika Hoepfner, Kristine Liao, and Geoffrey A. Ozin. 2014. Colloidal synthesis of 1T-WS$_2$ and 2H-WS$_2$ nanosheets: applications for photocatalytic hydrogen evolution. *Journal of the American Chemical Society* 136 (40):14121–14127.

Mahvelati-Shamsabadi, Tahereh, Hossein Fattahimoghaddam, Byeong-Kyu Lee, Hongsun Ryu, and J. I. Jang. 2021. Caesium sites coordinated in boron-doped porous and wrinkled graphitic carbon nitride nanosheets for efficient charge carrier separation and transfer: photocatalytic H$_2$ and H$_2$O$_2$ production. *Chemical Engineering Journal* 423:130067.

Matos, J. 2016. Eco-friendly heterogeneous photocatalysis on biochar-based materials under solar irradiation. *Topics in Catalysis* 59 (2):394–402.

Mian, Md Manik, and Guijian Liu. 2018. Recent progress in biochar-supported photocatalysts: synthesis, role of biochar, and applications. *RSC Advances* 8 (26):14237–14248.

Murali, G., Jeevan Kumar Reddy Modigunta, Young Ho Park, et al. 2022. A review on MXene synthesis, stability, and photocatalytic applications. *ACS Nano* 16 (9):13370–13429.

Nishiyama, Hiroshi, Taro Yamada, Mamiko Nakabayashi, et al. 2021. Photocatalytic solar hydrogen production from water on a 100-m^2 scale. *Nature* 598 (7880):304–307.

Norouzi, Omid, Amanj Kheradmand, Yijiao Jiang, Francesco Di Maria, and Ondrej Masek. 2019. Superior activity of metal oxide biochar composite in hydrogen evolution under artificial solar irradiation: a promising alternative to conventional metal-based photocatalysts. *International Journal of Hydrogen Energy* 44 (54):28698–28708.

Padma, Tatiparti, Dheeraj Kumar Gara, Amara Nadha Reddy, Surya Veerendra Prabhakar Vattikuti, and Christian M. Julien. 2022. MoSe$_2$-WS$_2$ nanostructure for an efficient hydrogen generation under white light LED irradiation. *Nanomaterials* 12 (7):1160.

Praus, Petr, Lenka Řeháčková, Jakub Čížek, et al. 2022. Synthesis of vacant graphitic carbon nitride in argon atmosphere and its utilization for photocatalytic hydrogen generation. *Scientific Reports* 12 (1):13622.

Qian, Junchao, Zhigang Chen, Hui Sun, et al. 2018. Enhanced photocatalytic H$_2$ production on three-dimensional porous CeO$_2$/carbon nanostructure. *ACS Sustainable Chemistry & Engineering* 6 (8):9691–9698.

Rao Pala, Laxmi Prasad, and Nageswara Rao Peela. 2021. Green hydrogen production in an optofluidic planar microreactor via photocatalytic water splitting under visible/simulated sunlight irradiation. *Energy & Fuels* 35 (23):19737–19747.

Saleem, Zubia, Erum Pervaiz, M. Usman Yousaf, and M. Bilal Khan Niazi. 2020. Two-dimensional materials and composites as potential water splitting photocatalysts: a review. 10 (4):464.

Sherryna, Areen, and Muhammad Tahir. 2021. Role of Ti$_3$C$_2$ MXene as prominent schottky barriers in driving hydrogen production through photoinduced water splitting: a comprehensive review. *ACS Applied Energy Materials* 4 (11):11982–12006.

Shu, Haibo, Dong Zhou, Feng Li, Dan Cao, and Xiaoshuang Chen. 2017. Defect engineering in MoSe$_2$ for the hydrogen evolution reaction: from point defects to edges. *ACS Applied Materials & Interfaces* 9 (49):42688–42698.

Song, Qi, Shiliang Heng, Wenbin Wang, Huili Guo, Haiyan Li, and Dongbin Dang. 2022. Binary type-II heterojunction K$_7$HNb$_6$O$_{19}$/g-C$_3$N$_4$: an effective photocatalyst for hydrogen evolution without a co-catalyst. *Nanomaterials (Basel)* 12 (5):849.

Sutar, Shubham, Sachin Otari, and Jyoti Jadhav. 2022. Biochar based photocatalyst for degradation of organic aqueous waste: a review. *Chemosphere* 287:132200.

Tahir, Muhammad, Areen Sherryna, Rehan Mansoor, Azmat Ali Khan, Sehar Tasleem, and Beenish Tahir. 2022. Titanium carbide MXene nanostructures as catalysts and cocatalysts for photocatalytic fuel production: a review. *ACS Applied Nano Materials* 5 (1):18–54.

Tang, Rui, Ruiyang Yin, Shujie Zhou, et al. 2017. Layered MoS$_2$ coupled MOFs-derived dual-phase TiO$_2$ for enhanced photoelectrochemical performance. *Journal of Materials Chemistry A* 5 (10):4962–4971.

Wan, Xuejuan, Haoqi Ke, Guanghui Yang, and Jiaoning Tang. 2018. Carboxyl-modified hierarchical wrinkled mesoporous silica supported TiO$_2$ nanocomposite particles with excellent photocatalytic performances. *Progress in Natural Science: Materials International* 28 (6):683–688.

Wang, Qing Hua, Kourosh Kalantar-Zadeh, Andras Kis, Jonathan N. Coleman, and Michael S. Strano. 2012. Electronics and optoelectronics of two-dimensional transition metal dichalcogenides. *Nature Nanotechnology* 7 (11):699–712.

Wang, Xinchen, Kazuhiko Maeda, Arne Thomas, et al. 2009. A metal-free polymeric photocatalyst for hydrogen production from water under visible light. *Nature Materials* 8 (1):76–80.

Wang, Yajun, Juan Chen, Liming Liu, et al. 2019. Novel metal doped carbon quantum dots/CdS composites for efficient photocatalytic hydrogen evolution. *Nanoscale* 11 (4):1618–1625.

Wang, Ying-Ying, Yan-Xin Chen, Tarek Barakat, et al. 2022. Recent advances in non-metal doped titania for solar-driven photocatalytic/photoelectrochemical water-splitting. *Journal of Energy Chemistry* 66:529–559.

Wang, Yingying, Yan-Xin Chen, Tarek Barakat, et al. 2021. Synergistic effects of carbon doping and coating of TiO$_2$ with exceptional photocurrent enhancement for high performance H$_2$ production from water splitting. *Journal of Energy Chemistry* 56:141–151.

Wang, Yiou, Anastasia Vogel, Michael Sachs, et al. 2019. Current understanding and challenges of solar-driven hydrogen generation using polymeric photocatalysts. *Nature Energy* 4 (9):746–760.

Wang, Yuanpeng, Xuqiang Hao, Lijun Zhang, Zhiliang Jin, and Tiansheng Zhao. 2021. Amorphous Co$_3$S$_4$ nanoparticle-modified tubular g-C$_3$N$_4$ forms step-scheme heterojunctions for photocatalytic hydrogen production. *Catalysis Science & Technology* 11 (3):943–955.

Wang, Zheng, Can Li, and Kazunari Domen. 2019. Recent developments in heterogeneous photocatalysts for solar-driven overall water splitting. *Chemical Society Reviews* 48 (7):2109–2125.

Wei, Lin, Deqian Zeng, Zongzhuo Xie, et al. 2021. NiO nanosheets coupled with CdS nanorods as 2D/1D heterojunction for improved photocatalytic hydrogen evolution. *Frontiers in Chemistry* 9:655583.

Wilson, J. A., and A. D. Yoffe. 1969. The transition metal dichalcogenides discussion and interpretation of the observed optical, electrical and structural properties. *Advances in Physics* 18 (73):193–335.

Wu, Haisu, Tifang Miao, Qinghua Deng, et al. 2022. Accelerating nickel-based molecular construction via DFT guidance for advanced photocatalytic hydrogen production. *ACS Applied Materials & Interfaces* 14 (15):17486–17499.

Wu, Youlin, Pengfei Zhu, Youji Li, Lijun Zhang, and Zhiliang Jin. 2022. In situ derivatization of NiAl-LDH/NiS a p–n heterojunction for efficient photocatalytic hydrogen evolution. *ACS Applied Energy Materials* 5 (7):8157–8168.

Xiao, Xudong, Siying Lin, Liping Zhang, et al. 2022. Constructing Pd-N interactions in Pd/g-C_3N_4 to improve the charge dynamics for efficient photocatalytic hydrogen evolution. *Nano Research* 15 (4):2928–2934.

Xiong, Ting, Wanglai Cen, Yuxin Zhang, and Fan Dong. 2016. Bridging the g-C_3N_4 interlayers for enhanced photocatalysis. *ACS Catalysis* 6 (4):2462–2472.

Xu, Li, Yuhui Tian, Tiefeng Liu, et al. 2018. α-Fe2O3 nanoplates with superior electrochemical performance for lithium-ion batteries. *Green Energy & Environment* 3 (2):156–162.

Yan, Hongjian. 2012. Soft-templating synthesis of mesoporous graphitic carbon nitride with enhanced photocatalytic H_2 evolution under visible light. *Chemical Communications* 48 (28):3430–3432.

Yang, Can, Beatriz Chiyin Ma, Linzhu Zhang, et al. 2016. Molecular engineering of conjugated polybenzothiadiazoles for enhanced hydrogen production by photosynthesis. 55 (32):9202–9206.

Yang, Lin, Shaohui Guo, and Xuanhua Li. 2017. Au nanoparticles@MoS_2 core-shell structures with moderate MoS_2 coverage for efficient photocatalytic water splitting. *Journal of Alloys and Compounds* 706:82–88.

Ye, Chen, Jia-Xin Li, Zhi-Jun Li, et al. 2015. Enhanced driving force and charge separation efficiency of protonated g-C_3N_4 for photocatalytic O_2 evolution. *ACS Catalysis* 5 (11):6973–6979.

Yu, Yaoguang, Gang Chen, Yansong Zhou, and Zhonghui Han. 2015. Recent advances in rare-earth elements modification of inorganic semiconductor-based photocatalysts for efficient solar energy conversion: a review. *Journal of Rare Earths* 33 (5):453–462.

Yuan, Chunfeng, Yongli Shen, Chunyu Zhu, et al. 2022. Ru single-atom decorated black TiO_2 nanosheets for efficient solar-driven hydrogen production. *ACS Sustainable Chemistry & Engineering* 10 (31):10311–10317.

Yuan, Hui, Fenjian Fang, Jing Dong, Weiwei Xia, Xianghua Zeng, and Wenfeng Shangguan. 2022. Enhanced photocatalytic hydrogen production based on laminated MoS_2/g-C_3N_4 photocatalysts. *Colloids and Surfaces A: Physicochemical and Engineering Aspects* 641:128575.

Yuan, Huiyu, Kai Han, David Dubbink, Guido Mul, and Johan E. ten Elshof. 2017. Modulating the external facets of functional nanocrystals enabled by two-dimensional oxide crystal templates. *ACS Catalysis* 7 (10):6858–6863.

Yuan, Yu-Peng, Li-Sha Yin, Shao-Wen Cao, et al. 2014. Microwave-assisted heating synthesis: a general and rapid strategy for large-scale production of highly crystalline g-C_3N_4 with enhanced photocatalytic H_2 production. *Green Chemistry* 16 (11):4663–4668.

Zhang, Guigang, Zhi-An Lan, and Xinchen Wang. 2017. Surface engineering of graphitic carbon nitride polymers with cocatalysts for photocatalytic overall water splitting. *Chemical Science* 8 (8):5261–5274.

Zhang, Jian-Hua, Mei-Juan Wei, Zhang-Wen Wei, Mei Pan, and Cheng-Yong Su. 2020. Ultrathin graphitic carbon nitride nanosheets for photocatalytic hydrogen evolution. *ACS Applied Nano Materials* 3 (2):1010–1018.

Zhang, Luhong, Zhengyuan Jin, Shaolong Huang, et al. 2019. Bio-inspired carbon doped graphitic carbon nitride with booming photocatalytic hydrogen evolution. *Applied Catalysis B: Environmental* 246:61–71.

Zhang, Xiao-Hong, Nan Li, Jiaojiao Wu, Yan-Zhen Zheng, and Xia Tao. 2018. Defect-rich O-incorporated 1T-MoS$_2$ nanosheets for remarkably enhanced visible-light photocatalytic H$_2$ evolution over CdS: the impact of enriched defects. *Applied Catalysis B: Environmental* 229:227–236.

Zhang, Zhao, Luhua Lu, Zaozao Lv, et al. 2018. Porous carbon nitride with defect mediated interfacial oxidation for improving visible light photocatalytic hydrogen evolution. *Applied Catalysis B: Environmental* 232:384–390.

Zhang, Zhaofu, Qingkai Qian, Baikui Li, and Kevin J. Chen. 2018. Interface engineering of monolayer MoS$_2$/GaN hybrid heterostructure: modified band alignment for photocatalytic water splitting application by nitridation treatment. *ACS Applied Materials & Interfaces* 10 (20):17419–17426.

Zhou, Jing-Wen, Xia Jiang, Yan-Xin Chen, Shi-Wei Lin, and Can-Zhong Lu. 2022. N, P self-doped porous carbon material derived from lotus pollen for highly efficient ethanol-water mixtures photocatalytic hydrogen production. *Nanomaterials* 12 (10):1744.

Zhou, Min, Li Li, Sai Zhang, et al. 2021. Surface engineering of 2D carbon nitride with cobalt sulfide cocatalyst for enhanced photocatalytic hydrogen evolution. 218 (10):2100012.

Zhou, Yunlong, Meng Sun, Teng Yu, and Jian Wang. 2022. 3D g-C$_3$N$_4$/WO$_3$/biochar/Cu^{2+}-doped carbon spheres composites: Synthesis and visible-light-driven photocatalytic hydrogen production. *Materials Today Communications* 30:103084.

Zhou, Zhixin, Jianhai Wang, Jiachao Yu, et al. 2015. Dissolution and liquid crystals phase of 2D polymeric carbon nitride. *Journal of the American Chemical Society* 137 (6):2179–2182.

4 2D Nanomaterials for Wastewater Treatment by Adsorption Strategies

Sona Stanly and Jith C. Janardhanan

4.1 INTRODUCTION

Water is essential for the existence of life on earth and the sustainability, and purity of water is inevitable. The linear expansion of the human population over years will face the scarcity of water very shortly. Of the total percentage, only 2.5% of water is available for the processing of human needs. This availability of clean water is further diminishing due to water pollution because of the easily susceptible nature of water to pollutants. Rapid industrialization and urbanization that are exponentially increasing along with the expanded human population are the main roots of water pollution and cannot be eliminated up to a level because of their mutual interconnection with the expansion of human life. Among various sources, industrialization is considered the major water pollution contributor. We cannot blame a particular industry as the main root cause of water pollution since almost all the industry pays for their contribution. Along with industrial waste, other sources such as agricultural waste, electronic waste, marine dumping, refineries, radioactive waste, global warming, oil spillage, etc. also contribute to water pollution.

The increased water pollution has demanded the emergence of methodologies for the remediation of water pollution. Physical water treatment, chemical water treatment, biological water treatment, sludge treatment, etc. are commonly employed for wastewater treatment. Precipitation, coagulation, adsorption, ion exchange, membrane filtration, reverse osmosis, electrodialysis membrane treatment, catalysis (both photo and electrocatalysis), magnetic separation, phytoremediation, etc. are employed for wastewater treatment (Sharma and Bhattacharya 2017). Among various methodologies, chemical water treatment has to be emphasized because diverse novel and efficient methods have been successfully implemented for the removal of pollutants from contaminated water. Owing to their efficiency, ease of operation, and ability to perform for both small- and large-scale applications, adsorption has emerged as a propitious platform for the decontamination of wastewater by removing pollutants (Rashid et al. 2021).

The adsorption process is a surface phenomenon involving both physisorption and chemisorption (Das et al. 2017). Physisorption involves physical interactions, more specifically involving van der Waals' interactions while chemisorption involves covalent forces of attraction originating from the electrostatic force of attraction between oppositely charged species (Figure 4.1). Microporous materials such as activated

DOI: 10.1201/9781003343899-4

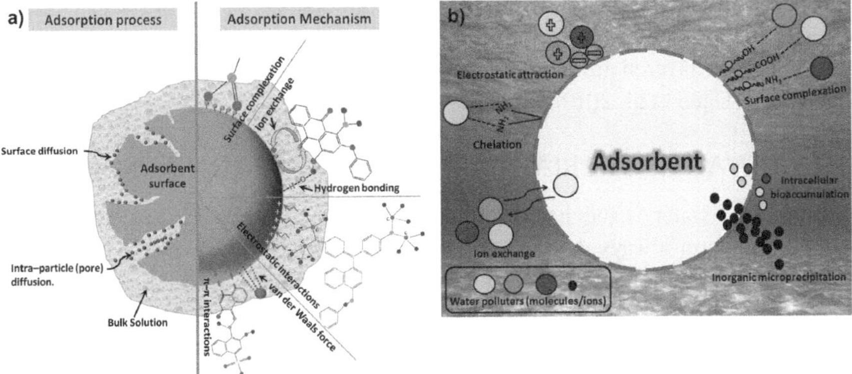

FIGURE 4.1 General schematic representation for the adsorption of organic dyes (a), metal-ions (b) from polluted water. Reproduced with permission from (Elwakeel et al. 2020; Dutta et al. 2021).

carbon, zeolite, alumina, silica, etc. are utilized as adsorbents for removing pollutants from wastewater. Since adsorption involves surface interactions, materials with a large effective surface area will contribute much to the adsorption phenomenon. In this context two-dimensional (2D) nanomaterial materials with highly exposed surfaces having maximized surface functionalities are the best fit for the purpose. Ultrathin nanosheets with large effective surface area, high area-to-thickness aspect ratio, more exposed surface with abundant active surface functionalities, widened interlayer distance have made 2D materials the best candidate for adsorption applications, especially for wastewater purification where hydrophilicity of 2D material is an important parameter (Fatima et al. 2022).

Graphene oxide (GO), molybdenum disulfide (MoS_2), MXenes, metal organic frameworks (MOF), clays, layered double hydroxides (LDHs), etc. are proven to be promising adsorbents in wastewater treatment. GO is a derivative of graphene with one atom thickness having oxygen-rich functionalities. The oxygenated functionalities not only impart hydrophilicity to GO but also will participate in the removal of organic/heavy metal ions (HMIs) from polluted water (Ahmad et al. 2020). MoS_2 belongs to the transition metal dichalcogenides family. Widen interlayer distance, defects, and sulfur-mediated interactions especially the Lewis soft-soft interaction between HMIs made MoS_2 an excellent adsorbent for removing HMIs from contaminated water (Ishag and Sun 2021). MXenes are transition metal carbides/nitrides and like GO, MXenes are hydrophilic owing to the abundant oxygenated surface functionalities. In addition, the hydroxyl-rich surface of MXene with a large surface area is proven to be an excellent adsorbent for water purification (Jeon et al. 2020). MOFs are porous crystalline materials with tunable porosity and for adsorption of pollutants from contaminated water, 2D-MOF nanosheets with coordinatively unsaturated metal centers, conductivity, large surface area, ease of post-functionalization are of best choice (Liu et al. 2021). 2D nanomaterials such as 2:1 clay (montmorillonite [MMT] clay), LDH, etc are also potential adsorbents because they have large surface areas that are accessible for pollutant adsorption. The clays' unmodified and modified forms are effective adsorbents for pollutant removal from water and are widely accepted due to their ease of use and economic viability (Peng et al. 2016). LDH is another layered anionic clay that is exploited as an adsorbent for anionic dyes from water (Daud et al. 2019).

4.2 2D-NANO MATERIALS FOR HEAVY METAL-ION ADSORPTION

The purity of water is inevitable for sustainable human life. But the rapid growth of the population always demands expanded industrialization which exponentially increases water pollution by charging highly toxic effluents into the water streams or sources. Among various organic and inorganic toxic water contaminants, HMIs are fatal because of their non-biodegradable, bioaccumulation nature and long-term exposure can bring severe health issues (Elwakeel et al. 2020). The tanning industry, batteries, metallurgical and chemical industries, refineries, electronic waste, etc. are the main culprits as the source of HMI-mediated water pollution. The peculiarity of HMI contamination is that even very low concentration of HMIs (in ppm) is sufficient to threaten the healthy life of humans. Cr, Ni, Hg, Pb, Zn, Cd, Ag, As, Mn, Cu, etc. are some of the main toxic HMIs effluents that are present in wastewater at different concentration levels. The health organization committees have implemented permissible levels for HMIs in water and above which will be fatal in terms of a healthy environment and human life (Muhammad Ekramul Mahmud, Huq, and Yahya 2016). For example, according to Environmental Protection Agency, the permissible levels of As(III), Pb(II), Cd(II), Cr(VI), Hg(II), Zn(II), Cu (II) are 0.01, 0.015, 0.005, 0.05, 0.002, 5.0, 1.3 ppm, respectively, for drinking water. Apart from drinking water, HMIs are also finding their way to the human body in the form of HMIs accumulated in fish plants, vegetables, etc. HMIs can bring severe health concerns to human life and it is vital to remove HMIs from water for healthy sustainable human life.

Because of the adverse effects of HMIs on human health, various methodologies and operating procedures have been developed for the decontamination of water by removing HMI effluents. Precipitation, oxidation, membrane filtration, reverse osmosis, adsorption, coagulation, etc. are the successfully employed methods for the removal of HMIs from water (Sharma and Bhattacharya 2017). Among these, adsorption has attracted great interest because of its cost-effective nature, ease of action, and modulating scalable nature for both small and large scales. Various adsorbents such as activated carbon, zeolites, clay, MOF/covalent organic framework, LDHs, etc. are widely explored for water decontamination by HMI adsorption. Owing to their excellent adsorption properties, among different adsorbents, 2D-layered materials have gained great attention (Fu et al. 2018). The superior adsorption properties can be attributed to the large effective surface area and high area-thickness aspect ratio. Besides, the electron-rich surface functionalities such as oxygenated, amino, imino functional moieties on the surface of nanosheets facilitated the covalent/ non-covalent/electrostatic interaction with HMIs. On account of these, more specifically, 2D layers with highly exposed surfaces having augmented concentration of active sites ensure the HMIs adsorption. 2D materials such as molybdenum disulfide, GO, MXenes, MOFs, etc. are successfully employed for the adsorption of HMIs for the treatment of wastewater.

4.2.1 MoS$_2$-Based Material for HMI Decontamination

MoS$_2$, possessing a layered structure comprising a sandwiched layer of Mo atom between two sulfur atoms layer through strong covalent bonds, is the most studied transition metal dichalcogenide, next to graphene (Naik and Rabinal 2020). The strongly covalent bounded monolayers of MoS$_2$ are stacked together by weak van der Waals' interaction with an interlayer distance of 0.65 Å and the weak interaction between the layers facilitates the interlayer expansion engineering which will be helpful for the adsorption of HMIs (Gupta, Chauhan, and Kumar 2020). Besides, the unique anisotropy of 2D morphology of sulfur-rich MoS$_2$ provides a large effective surface area and ion-permeable channels for HMIs adsorption. The rate of diffusion of high valence metal ions of large ionic radius can be accelerated with the widening of MoS$_2$ interlayers by reducing the ion-diffusional resistance. Besides, the interlayer expansion also augmented the surface exposed sulfur atom binding sites for ion adsorption. Furthermore, in addition to the hydrophilic nature and widen interlayer distance, the degree of defect in MoS$_2$ also plays a crucial role in the adsorption of HMIs from wastewater.

Interlayer widening which hinders the structure rupturing during the HMI insertion into the layer is the best strategy to adopt for the decontamination of wastewater by adsorption of HMIs. The interlayer widening is successfully employed for the removal of Cr(VI) which is considered as highly toxic mobile carcinogenic commonly found in industrial effluents. Defect-rich ultrathin nanosheets with an expanded interlayer distance of 9.5 Å exhibited better Cr(VI) removal with Cr(VI) intake capacity of 84.03 mg L^{-1} in Langmuir adsorption simulation where holes and cracks in the defect-rich nanosheets provide Cr(VI) diffusion and transport channels (Sun et al. 2018). Cr(VI) removal is found to be influenced by the synergism of both direct and reductive indirect adsorption where the latter involves the electrostatic driven adsorption of Cr(VI) on to negatively charged MOS$_2$ surface while the former consists of the reduction of a fraction of Cr (VI) into low toxic Cr(III) during the adsorption of Cr(VI) followed by subsequent adsorption of Cr(III) by ultrathin MoS$_2$ nanosheets. Wang and co-workers have found that IT-MoS$_2$ nanosheets with a widened interlayer distance of 9.4 Å displayed excellent sorption capacity of 200.3 mg g^{-1} for Cr (VI) ions and the superior sorption intake capacity can be attributed to co-ordinated Mo(IV) coordination sites which promote Cr(VI) to Cr(III) reduction during the adsorption of Cr(VI) ions (Li, Fan, et al. 2020). By exploiting the soft-soft interaction between sulfur and Hg atoms, Lu and co-workers demonstrated the removal of Hg(II) by defect-rich MoS$_2$ nanosheets (W-DR-N-MoS$_2$) having widened interlayer distance of 9.42 Å with near theoretical Hg(II)-uptake capacity of 2506 mg g^{-1} (Ai et al. 2016). The excellent Hg^{2+}-adsorption characteristics of W-DR-N-MoS$_2$ can be attributed to maximized exposed sulfur binding atoms originating from the widening of interlayer distance and defect induced additional mass transport channels in addition to a small particle size which also augmented the abundance of exposed sulfur binding atoms for Hg^{2+}. A few layer MoS$_2$ comprising large exposed sulfur atoms can act as a super adsorbent for the removal of Hg^{2+} ions from water (Jia et al. 2017). Exfoliated MoS$_2$ nanosheets with less than five layers exhibited a superior Hg^{2+}-uptake capacity of 800 mg g^{-1} and the excellent adsorption property can directly be assigned to multi-layer adsorption

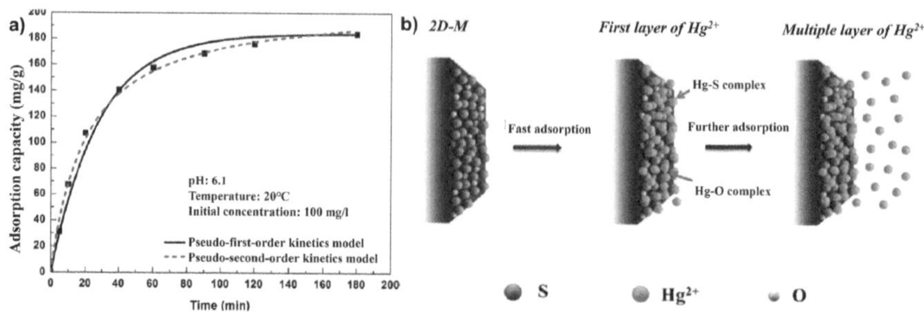

FIGURE 4.2 Hg^{2+}-adsorption kinetics of exfoliated MoS_2 nanosheets (a) and mechanism for adsorption. Reproduced with permission from (Jia et al. 2017).

(Figure 4.2). In other words, the first adsorption can be attributed to S-Hg, O-Hg complexation where O was formed in the nanosheets by partial oxidation of Mo and the second mode of adsorption originated from the electrostatic interaction between negatively charged MoS_2 surface and positively charged Hg^{2+}. Ray, Kumar and co-workers utilized morphology tuning accompanied by interlayer widening for the development of MoS_2 based super Pb(II)-selective adsorbents (Kumar, Fosso-Kankeu, and Ray 2019). The morphology tuning was achieved with sodium diethyldithiocarbamate trihydrate (DDC) and ethylenediaminetetraacetic acid (EDTA) at different pH and among different morphologies (such as microspheres, microrods, microrods with hair-like structures (MoS_2-N-H)), MoS_2-N-H with an expanded interlayer distance of 10 Å displayed selective Pb(II) adsorption with a highest capacity of 303.04 mg g^{-1}. Furthermore, MoS_2-N-H can adsorb Pb(II) selectively from real mine water with 100% efficiency. The widened interlayer distance of MoS_2-N-H provides enormous surface exposed S-sites for adsorption which is reflected in the superior Pb(II) adsorption which was well fitted with Freundlich isotherm. In addition, the excellent adsorption can also be attributed to Lewis-base interaction which leads to Pb-S complexation and Pb(II) – DDC/EDTA electrostatic interactions (more specifically interaction between Pb(II) and C/N/O- moieties).

Luo et al. have identified that phase change in MoS_2 from semiconducting 2H-phase into metallic 1T-phase has unraveled the enormous adsorption properties of MoS_2 such as enhanced adsorption uptake capacity with fast kinetics toward Pb(II) and Cu(II) HMI adsorption (Luo et al. 2019). It was observed that compared to 2H-MoS_2, 1T-MoS_2 exhibited superior HMI adsorption capacity of 147.09, 82.13 mg g^{-1} for Pb(II) and Cu(II), respectively. On counting the indispensable contribution of S-binding atoms to HMI adsorption, the excellent adsorption properties of 1T-MoS_2 over 2H-MoS_2 can be assigned to the easy feasible transfer of near Fermi energy contributing electrons into the LUMO of 1T-MoS_2 because of the metallic character of 1T-MoS_2.

4.2.2 Graphene-Based Material for HMIs Decontamination

Graphene is the most studied 2D-layered material comprising a honeycomb lattice of sp^2-hybridized C-atoms arranged hexagonally with one-carbon atom thickness. High surface area, good electrical conductivity, and excellent thermal and mechanical stability have made graphene inevitable for application in diverse scientific areas (Allen, Tung, and Kaner 2010). Owing to its high surface area, graphene is also employed as a good adsorbent but the hydrophobic nature makes it more appropriate in oil phase applications than in the aqueous phase which hinders its application in wastewater treatment (Zhao et al. 2014). For wastewater treatment, GO, one C-atom thick derivative of graphene formed by oxidation followed by exfoliation can be employed (Velusamy et al. 2021). Apart from the sp^2-hybridized C-atoms in graphene, GO consists of sp^3-hybridized carbon atoms and oxygen-rich functionalities such as –OH, –COOH, C=O, and epoxide rings. The oxygen-rich surface functionalities hold lone pairs of electrons which impart an overall negative charge to the GO surface and can be effectively exploited for the adsorption of electron-deficient HMIs by utilizing electrostatic interactions. In addition, these surface functionalities can induce surface complexation with HMIs by utilizing a chelation mechanism. Similar to graphene, single-atom-thick GO possesses a large specific surface area which can be beneficial for the adsorption of HMIs. Besides, the oxygen-rich functionalities and imparted surface negative charge made GO hydrophilic and can well be dispersed in water which is an advantage for wastewater remediation. Though GO possesses structural and morphology advantages for HMI adsorption, the high hydrophilicity hinders the process because the high hydrophilicity results in soluble GO dispersion which makes it difficult to remove from the water after adsorption. The agglomeration tendency of GO after adsorption also accounts for the minimal usage of GO in its native form for HMIs adsorption applications. The bottleneck can be overcome with the hybridization of GO with other nanomaterials (Peng et al. 2017). The post-functionalization of GO can be made with ease of surface oxygen-rich functionalities. Chitosan, polypyrrole, polyaniline, polyacrylonitrile, etc. have been successfully hybridized with GO for efficient adsorption of HMIs. The hybridization modulates the hydrophilicity and also enhances the effective contact surface area which promotes the HMI affinity of GO surface.

Magnetic few-layered GO (FLMGO) comprising GO-Fe_3O_4 hybrid architecture have adopted the chelation-like chemical interactions of C=O, C–O moieties toward Cd(II), Cu(II) and exhibited quick adsorption of Cd(II), Cu(II) with uptake capacities of 401.14, 1142.22 mg g^{-1} in 5, 7 min, respectively (Guo et al. 2021). The coordination between Cd(II), Cu(II), and lone pair of electrons in oxygen functionalities of GO is very efficient which is reflected in their fast adsorption kinetics. Huang and co-workers have modulated the adsorption properties of GO for HMI adsorption by hybridizing with hyperbranched polyamide amine (HPAMAM) and microcrystalline cellulose (Liu et al. 2022). Covalent grafting of HPAMAM onto GO increases the specific surface area and surface-active groups for the adsorption phenomenon. Concurrently, the introduction of dialdehyde cellulose (DAC) which is the oxidized product of microcrystalline cellulose on to GO-HPAMAM creates micro/nano bumps which in turn enhance the contact angle and surface hydroxyl groups for better adsorption. Consequently, GO-HPAMAM-DAC hybrid displayed a better adsorption capacity of 680.3, 418.4, 280.1 mg g^{-1} for Pb(II), Cd(II), and Cu(II), respectively.

It was found that covalent functionalization of GO with cross-linked chitosan via amino propoxy silane (CHI-APSGO) enhances the adsorption properties of GO for selective Pb(II) removal by adsorption (Sharma, Singh, and Shahi 2019). Covalent grafting imparts large surface-active HMI-loving –NH_2/=NH moieties

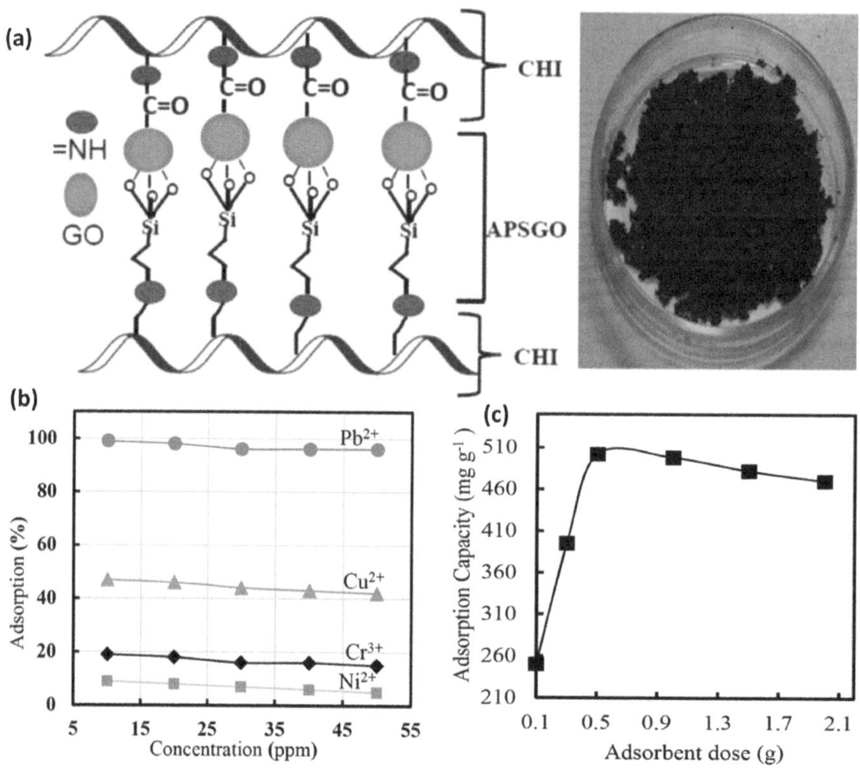

FIGURE 4.3 Schematic representation of CHI-APSGO (a), % of adsorption of Pb(II), Cr(III), Cu(II) and Ni(II) adsorption by CHI-APSGO (b) and adsorbent dose variation of Pb(II) (c). Reproduced with permission from (Sharma, Singh, and Shahi 2019).

which facilitate covalent bonding with metal ions concomitantly with the affinity of oxygen lone-pair of electrons to metal-ions (Figure 4.3). The surface active nitrogen and oxygen-rich functionalities with Pb(II)-ion affinity impart excellent adsorption properties to CHI-APSGO and displayed an uptake capacity of 566.2 mg g⁻¹ at pH of 5.0. Apart from chitosan, polyamines with abundant free amino groups are a good choice for hybridization with GO for HMI adsorption applications. GO was coated with polydopamine (PD) followed by covalent grafting with polyethylenimine with plenty of amine moieties (Dong et al. 2015). PD coating on GO enhances its affinity toward metal ions with contribution from abundant surface amine-based coordinating moieties. The influence of amine moieties favours the adsorption of HMIs such as Cu(II), Cd(II), Pb(II), Hg(II) with adsorption capacities of 87, 106, 197, 110 mg g⁻¹, respectively.

4.2.3 MXene-Based Material for HMIs Decontamination

MXenes are 2D transition metal carbides/nitrides with general formula $M_{n+1}C_nT_x$ where M is an early transition metal (such as Ti, Mo, Nb, etc.), C is carbon/nitrogen and T represents surface moieties such as F, –OH (Chen et al. 2022). The surface functionalities are imparted onto the MXene surface by the etching of a layer from the MAX phase by HF etching. The surface functionalities, especially hydroxyl groups, have a crucial impact on the adsorption properties of MXene in terms of electrostatic interactions and surface complexation (Othman, Mackey, and Mahmoud 2022). Besides, 2D MXene nanosheets exhibited augmented effective surface area, hydrophilicity, better water dispersibility, conductivity and are beneficial for HMI decontamination by preferential adsorption (Sheth et al. 2022).

Generally, titanium exhibited excellent HMIs adsorption properties and cation-substituted [Ti-OH]*-ion exchange site provides the selective sorption sites for HMIs. Considering the preferential sorption behavior of Ti-OH bonds, Ti-based MXene which was prepared by alkalization intercalation (alk-MXene; $(Ti_2C_2(OH/ONa)_xF_{2-x})$) exhibited preferential and quick sorption of Pb(II) in the presence of competing Mg(II) and Ca(II) ions (Peng et al. 2014). The layered structure with a large surface area comprising activated Ti-OH and inner sphere complexation contributed to the superior Pb(II) preferential sorption behavior where inner field complexation facilitated the $[Ti-O]-H^+-Pb(II)$ interaction (Figure 4.4). The high surface area with augmented hydroxyl groups of Ti_3C_2 MXenes was also exploited in Cr(VI) decontamination by adsorption. HF etched Ti_3C_2Tx Mxenes with surface hydroxyl groups exploited for water treatment and exhibited Cr(VI) adsorption uptake capacity of $250\,mg\,g^{-1}$ with residual Cr(VI) concentration of less than 5 ppm (Ying et al. 2015). Cr(VI) decontamination involves the adsorption of Cr(VI) facilitated by surface hydroxyl groups followed by its reduction to Cr(III) mediated by electron transfer from MXene which later can be removed simply by adjusting the pH to 5. Like GO, where oxygenated surface functionalities contribute much to adsorption behavior, the hydroxyl groups on MXenes which are imparted into layered structure by HF etching have a vital role in Cu(II) adsorption kinetics (Shahzad et al. 2017). Owing to excellent morphological and structural characteristics such as high surface area, hydrophilic terminal surface groups, better dispersibility in water-delaminated $Ti_3C_2T_x$ exhibited Cu(II) sorption properties with uptake adsorption capacity of $78.45\,mg\,g^{-1}$. The adsorption process is very fast and has achieved equilibrium within 3 min and proceeds via the formation of Cu_2O, CuO species by the interaction between surface terminated hydroxyl groups of MXene with Cu(II)-ions.

The introduction of the chelating group further enhances the adsorption behavior of MXenes and the surface functionalization of MXenes with chelating –COOH group enhances the adsorption uptake capacity (Isfahani et al. 2022). Without affecting the hydrophilicity, covalently grafted carboxylic acid groups facilitated the interlayer ion diffusivity and thereby enhanced the Hg(II) adsorption properties. Furthermore, the –COOH groups provide active sites for Hg(II) chemisorption and promote electrostatic interaction and surface complexation with the reduction of adsorbed Hg(II)-ions. These interactions are accountable for the superior Hg(II) adsorption behavior of MXenes-COOH which exhibited an adsorption uptake of $499.7\,mg\,g^{-1}$. Restacking of MXene layers is found to have a negative impact on the HMI adsorption properties of MXenes and Xu and co-workers have exploited reduced GO (rGO) to prevent the restacking of MXene layers for HMI adsorption

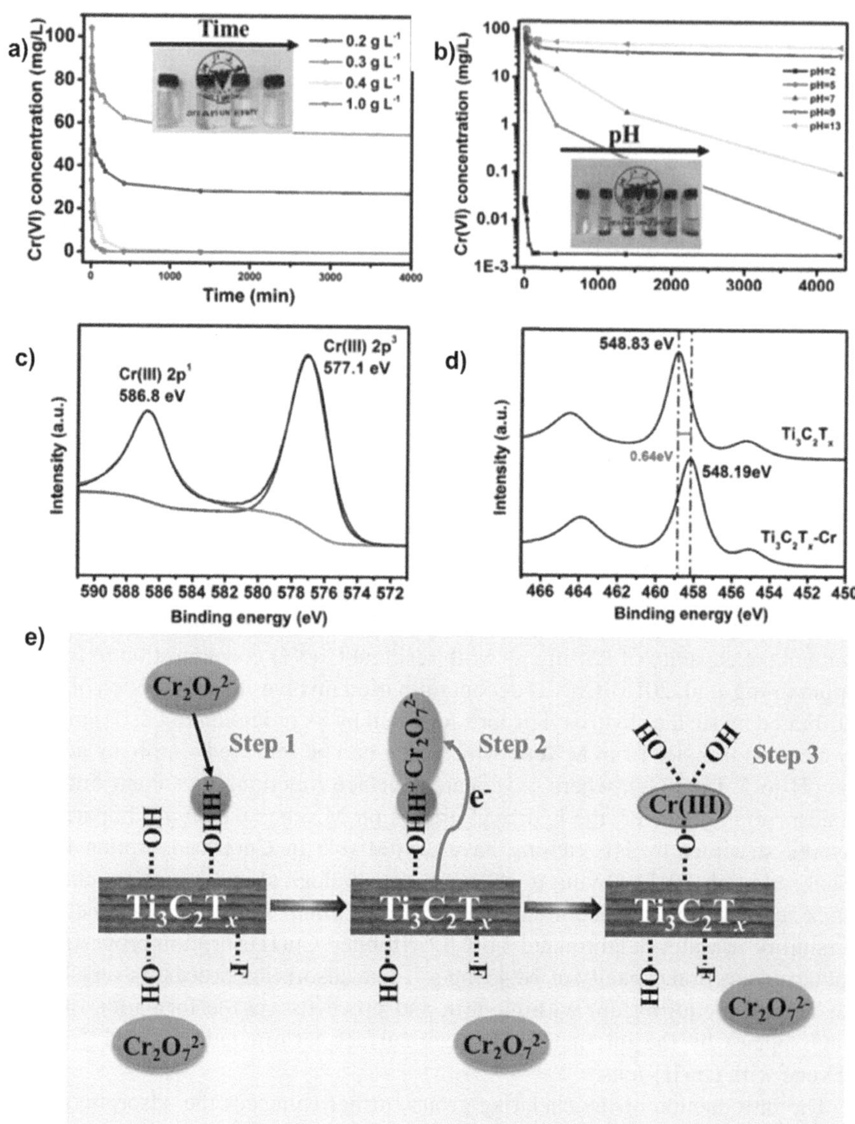

FIGURE 4.4 Dosage effect of MXene nanosheets on Cr(VI) removal (a), pH dependent removal of Cr(VI) ions (b), XPS spectrum of MXene nanosheets after Cr(VI) adsorption (c, d) and schematic representation for Cr(VI) removal by MXene nanosheets (e). Reproduced with permission from (Peng et al. 2014).

applications (Xie et al. 2019). Here, rGO enhances the MXene-HMI interaction by mitigating the restacking of MXene nanosheets. Further, the maximized surface hydroxyl groups formed by HCl surface modification favors the charge transfer from MXene nanosheets to HMIs which in turn facilitated their reduction. The synergism of both morphological and terminal surface hydroxyl groups thereby promotes the excellent adsorption toward Cr(VI), Pd(II), Au(III), and Ag(I) with adsorption uptake capacity of 84, 890, 1241, 1172 mg g^{-1}, respectively.

4.2.4 MOF-Based Material for HMIs Decontamination

MOFs are porous π-conjugated crystalline materials possessing metal nodes and organic chelating groups coordinated via coordination terminals of chelating ligands (Wang and Astruc 2020). The π-conjugated skeleton of MOF can be tuned with the selection of chelating ligands which in turn tunes both the porosity and morphology of MOFs. Tunable porous nature, electrical conductivity, active catalytic sites, etc. made MOF an interesting material for scientific applications, especially in the area of adsorption (Chen, Bai, and Ye 2020). In the case of adsorption, one hurdle in the form of a coordinatively unsaturated metal center can be overwhelmed by implementing bimetal or multi-metal nodes. Compared to 3D bulk MOF, 2D-MOF nanosheets possess thin morphology, highly exposed surface with active sites, high area-thickness aspect ratio, effective surface area and can be well applied in the wastewater treatment by adsorption of HMIs (Liu et al. 2021).

For adsorption-based applications, the selection of 2D-MOF precursors has a crucial role because other than structural building units, the additional functionalities such as –OH, –NH$_2$, =N–, etc. promotes the interaction with HMIs and thereby contribute to the adsorption properties. Zn-based 2D-MOF (Zn(Bim)(oAc)) nanosheets comprising benzimidazole (Bim) and acetate (oAc) ligands exhibited preferential Pb(II), Cu(II) adsorption with adsorption capacity of 253.8, 335.6 mg g^{-1} within 30–90 min (Xu et al. 2020). The excellent adsorption behavior of Zn(Bim)(OAc)-NS can be attributed to increased exposed active sites consisting of surface imino, hydroxyl groups, and ultrathin thickness of MOF nanosheets. The imino, hydroxyl groups have a vital role in the preferential adsorption properties where they acted as adsorption sites for Pb(II), Cu(II) and their strong binding ability facilitating the chemisorption of monolayers, resulted in the selective Pb(II), Cu(II) adsorption. Few layer sulfur-rich MOF (Co(CNS)$_2$(pyz)$_2$) nanosheets exhibited excellent HMI adsorption properties with good recyclability up to three times (Li, Duan, et al. 2020). Thio group mediated HMIs coordination-assisted chemisorption via inner sphere complex is responsible for the better selective adsorption behavior of CoCNSP nanosheets to HMIs. The thiol-assisted coordination is selective in the order of Co(II), Zn(II), Cd(II), Ni(II) < Cu(II) < Pb(II) < U(VI) < Hg(II) with adsorption capacities of 716, 661, 534, 325 mg g^{-1}, respectively, for Hg(II), U(VI), Pb(II), Cu(II)-ions. Liu, Fang and co-workers developed bimetallic coordinated Ni/Cd-MOF for selective Pb(II) adsorption by one-step hydrothermal method (Figure 4.5) (Li et al. 2021).

Bimetallic Ni/Cd-MOF comprises 4–5 layers with a large effective surface, porosity and the surface functionalities impart surface negative charge that exhibit high affinity toward Pb(II) ions. The crystal defects and vacancies originating from lattice distortion and unbalanced bimetallic coordination have played an inevitable role in the preferential adsorption of Pb(II). In addition, the uncoordinated surface –COOH groups act as binding sites for Pb(II)-ions and owing to the synergistic effects of both, Ni/Cd-MOF with more surface –COOH groups exhibited superior Pb(II) adsorption with uptake capacity of 950.61 mg g^{-1}.

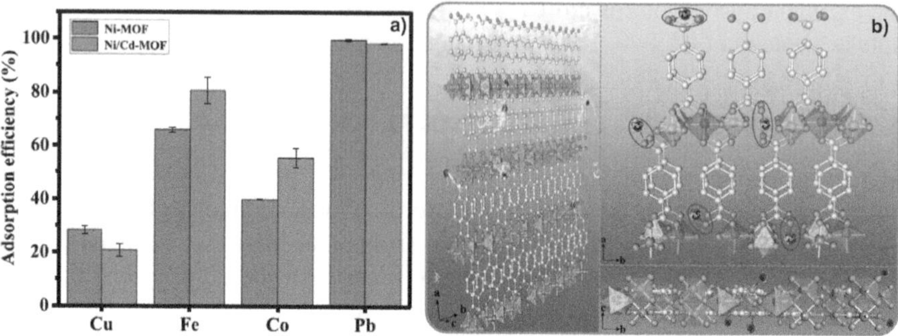

FIGURE 4.5 Adsorption of various HMIs (a) and mechanism for preferential Pb(II) adsorption by Ni/Cd-MOF. Reproduced with permission from (Li et al. 2021).

4.3 2D MATERIALS FOR DYE ADSORPTION

The world's half population is under water scarcity since the water bodies are polluted with dyes, heavy metals, and other organic pollutants due to intense population growth and rapid industrialization. One of the major threats is the color-imparting material such as organic dyes from the textile and dyeing units. Most of these materials are stable under normal environmental conditions and are harmful to aquatic life and the ecosystem (Sham and Notley 2018) due to their carcinogenic and mutagenic characteristics. There are different methods adopted to remove these materials from water which include adsorption, photodegradation, reverse osmosis, coagulation, membrane filtration, etc. The adsorption attained a great interest among all these methods because of its economic viability and ease of operation. Adsorption, one of the eco-friendly processes of removal of dyes from water, can be employed using various nanomaterials like carbon nanotubes, carbon nanofibers (CNF), graphene, different clays, LDH, etc.

Owing to their large surface area, 2D nanomaterials such as 2:1 clay (MMT clay), graphene, LDH, etc. are proven to be potential adsorbents because a large surface area is accessible for pollutant adsorption. The clays either in pristine or modified form are efficient and cost-effective adsorbents for pollutant removal (Peng et al. 2016). Among different clays, MMT clay, one of the soft-layered pyrosilicate clay, is extensively used because of its large surface area available for adsorption. The exchangeable cations are available in between the layers which could be used for modification of the clays which in turn improves the adsorption capacity. Graphene possesses excellent mechanical properties, flame resistance barrier properties, and electrical conductivity. Besides, it stands out as a good adsorbent for dyes, metals etc. As a potent dye adsorbent, the main advantage of graphene is its exceptionally high surface area ($2630\,m^2g^{-1}$) (Zhu et al. 2016; Chen et al. 2013). LDH is another layered anionic clay that is exploited as an adsorbent for anionic dyes from water (Daud et al. 2019).

4.3.1 NANO CLAYS FOR DYE REMOVAL

Natural and synthetic clays are used for the adsorption of various contaminants from water. Clays are naturally occurring phyllosilicate minerals with varying content of water trapped in the mineral structure having a particle size of less than 2 μm (He et al. 2014). Large surfaces and porosity nature are beneficial for physical and chemical reactions of various contaminant materials on the surface of the clay. These surface reactions on the surface of the clay are also increased due to its high electrostatic repulsion and specific cation exchange with crystallinity. The presence of cations such as Ca^{2+}, Mg^{2+}, and Na^+ exist in the interlayers or on the surfaces of the clay which are present to balance the charge, making the clays capable of dealing with the pollutants by ion-exchange methodology. These properties make it useful in adsorbing various pollutants from water including dyes, heavy metals, organic pollutants, etc. The larger the porosity on the surface of nano-clay, the higher will be its bonding power on the surface with containment (Awasthi, Jadhao, and Kumari 2019).

The clays are classified based on the layered structure they possess. They are 1:1 clay which is also known as the kaolinite group and have a strong van der Waals force of attraction between the layers. The second one is 2:1 and called as smectite group which is having 2:1 crystal lattice with various isomorphous substitutions to generate permanent negative charges on the surface and balancing cations in the interlayers. The next group is 2:1:1 which has a basic 2:1 structure. It has a general structure of one or two tetrahedral silica sheets attached to an octahedral aluminium sheet, which are bonded together mainly in a 2:1 (T-O-T) and 1:1 (T-O) ratio. $Si_2O_6(OH)_4$ units are the basic building unit of tetrahedral silica sheet and the octahedral aluminium sheet is mainly composed of $Al_2(OH)_6$ units. Smectite has 2:1 crystal lattice with various isomorphous substitutions that generate permanent negative charges on the surface and balancing cations in the interlayers.

All clays are used as a potential adsorbent in water purification but the 2:1 clay has got great attention due to the availability of exchangeable cations and which are 2D materials with a planar structure comprising good surface area with porosity. 2D clays such as MMT/bentonite/smectite, kaolinite, illite, sepiolite, palygorskite/attapulgite, halloysite, etc. exhibited excellent efficiency for the removal of colored materials from water. They possess charge on their surface due to the isomorphous substitution along with counter cations in between the layers to balance the negative charges (Zhu et al. 2016; Belhouchat, Zaghouane-Boudiaf, and Viseras 2017). Among the 2D clays, MMT clay is considered a good adsorbent for dyes due to its economic viability, eco-friendly nature, ease of modification, and cost-effective nature compared to other carbonaceous materials. Owing to the presence of surface negative charges and cation exchange capacity, MMT clay has mainly been employed to remove cationic dyes (He et al. 2014).

The MMT clay has negative charges on the surface due to the isomorphous substitution of Al^{3+} for Si^{4+} in the tetrahedral layer and Mg^{2+} for Al^{3+} in the octahedral layer. To balance the surface charges, there are exchangeable cations present between the interlayer galleries. The MMT has an interlayer distance of 1.22–1.42 nm and consists of several hundred individual platelets with 1.0 nm thick, 50–100 nm lateral dimension, and 0.3 nm between two adjacent particles. The contaminants are adsorbed on the clay substrate through different mechanisms such as surface adsorption, partition, surface precipitation, and structural incorporation (Zhu et al. 2016). There are different physico-chemical techniques developed to remove the dyes, one of the potentially hazardous materials from effluent treatment. Adsorption is the most important mechanism in terms of cost, design, and dye separation efficiency when compared with other chemical treatments including photolysis and photocatalytic processes, anaerobic, aerobic degradation, and physicochemical methods including electrokinetic coagulation, ion exchange, adsorption, and membrane filtration, etc. (Kausar et al. 2018; Unuabonah et al. 2013; Sophia A and Lima 2018; Zhu et al. 2016).

Adsorption can be defined as the process in which the solute molecule (adsorbent) adheres to the surface of a solid or liquid substance known as the adsorbate and forms a thin layer on the surface. It is a mass transfer process that occurs through the solid adsorbent by removing a dissolved contaminant from an aqueous solution by attracting it toward the adsorbent surface (Kausar et al. 2018). The extent of adsorption depends mainly on the nature of the adsorbent. The efficiency of adsorption is mainly influenced by molecular size, molecular structure, and molecular weight. Besides solution concentration, polarity and adsorbent surface properties like surface area and particle size, etc. have also influenced the efficiency of adsorption. The contaminants can adhere to the surface by chemical, physical, or ion exchange mechanism till the available surface area is saturated after that regeneration/recycling is needed depending on its feasibility (Ngulube et al. 2017). In MMT clay, surface negative charges are highly influential to adsorb the cationic materials such as dyes while the ion exchange is more predominant in heavy metal adsorption. The acid treatment increases the – OH groups stoichiometry which is useful for heavy metal removal as well as anionic dye removal (Zhu et al. 2016). The net negative charges on the surface of nano clay provide sites for the adsorption of positively charged pollutants like cationic dyes and thereby increase the adsorption capacity. The exchangeable cation is the other part involved in the adsorption process without affecting the structure of the clay. The high surface area and porosity also act as key factors as it provides the sites for the adsorption. The adsorption capacity of these clays can be fine-tuned by modulating the surface area, porosity, and surface charges. As already discussed, the surface area is one of the key factors in pollutant removal by adsorption and which can be improved by surface modification. The modifiers enter into the interlayer galleries and the distance between layers enlarges. The sufficiently large interlayer spacing accommodates a high amount of dyes from the contaminated water. As shown in Figure 4.6, the CTAB (cetyltrimethylammonium bromide)-modified MMT helps in intercalating the layers and positively changes the adsorption capacity (Huang et al. 2017).

4.3.2　Graphene-Based Material for Dye Removal

Graphene in the form of a single-layer sheet was developed by Geim and Novoselov in 2004. Graphene is the thinnest and lightest material known today with the highest strength and Young's modulus. It is a good conductor of heat and electricity at room temperature. Graphene is an allotrope of carbon, one atomic thick sp^2-hybridized monolayer form with a planar structure and carbon atoms tightly arranged in a hexagonal honeycomb lattice. By introducing oxygen-containing functional groups such as carboxylic groups (–COOH), hydroxyl (–C–OH), epoxy groups (–O–), carbonyl (–C–O) groups, etc. on the edges of the surface of graphene, the sp^2-hybridized carbon framework transformed into a sp^3 which will be favorable for interaction with both positive and negative specious. All of these active sites available at the edges are

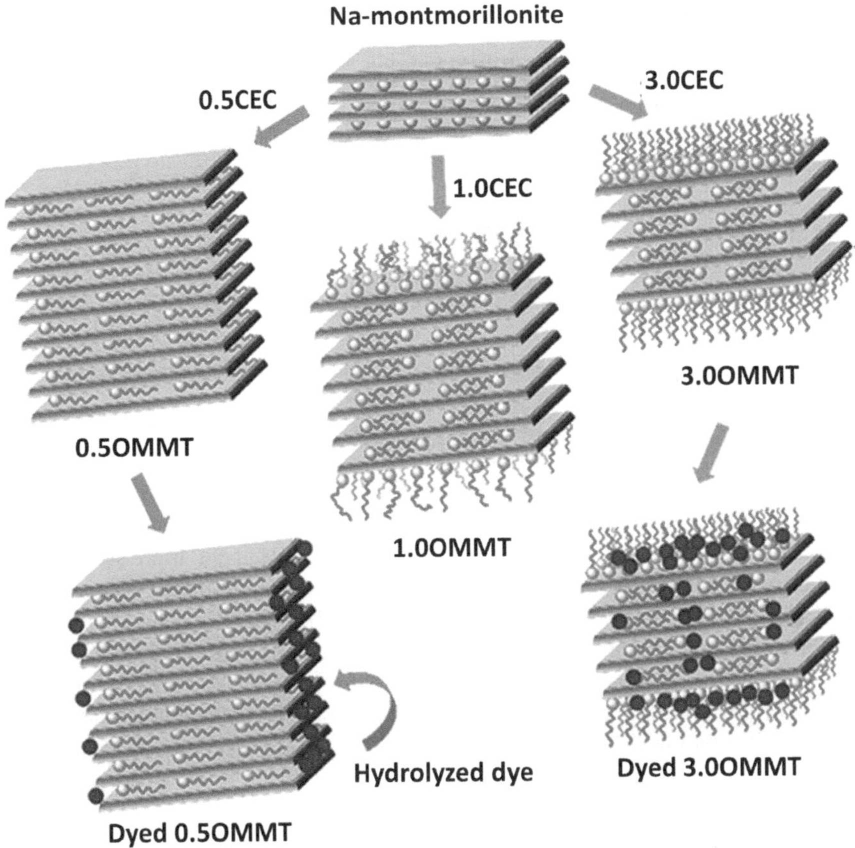

FIGURE 4.6 Modified MMT-based material for contaminant removal. Reproduced with permission from (Huang et al. 2017).

highly suited to adsorption applications (Abu-Nada, Abdala, and McKay 2021). It is one of the best materials for environmental remediation for the removal of organic dyes, metal ions, oils from water, and poisonous gases from the atmosphere (Ye, Liu, and Feng 2017; Aboelfetoh, Elhelaly, and Gemeay 2018; Saleh et al. 2013; Zhang et al. 2016; Nováček et al. 2017).

Graphene synthesized by chemical method is widely applied for the adsorption of various pollutants from water. The different oxygen-containing polar groups can be introduced at the basal plane of the graphene sheet by the chemical oxidation method which will also increase the interplanar distances between the graphene layers and thereby reducing the π–π interaction between the layers and is the most convenient method to prepare exfoliated GO. (Sham and Notley 2018; Perreault, Fonseca de Faria, and Elimelech 2015). The complete removal of those polar groups and restoring the graphitic planes after the reduction of GO formed by the chemical method is a major challenge. Hence, the obtained graphene is more previously named as rGO. The unreduced oxygen functionalities attached in rGO impart hydrophilicity and these

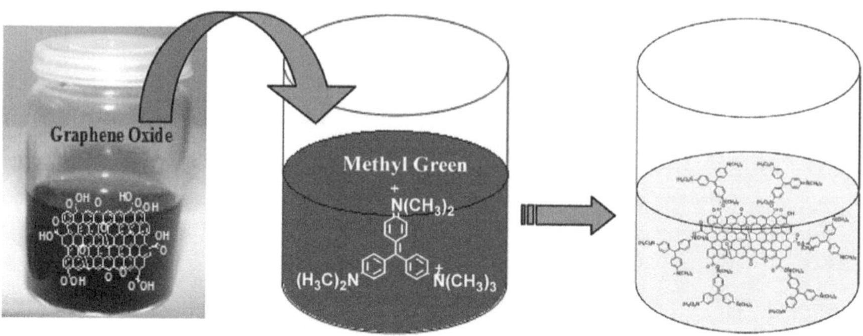

FIGURE 4.7   GO as a potential dye adsorbent. Reproduced with permission from (Sharma and Das 2013).

groups can act as the site for reaction with the organic pollutants (Sharma and Das 2013). Even though most of the carbonaceous materials are good in the adsorption of poisonous materials including dyes from water due to its easy adsorption-desorption phenomenon, the exceptional properties especially extremely high surface area $(2630\,m^2g^{-1})$, porosity, etc. improve the potential adsorption capacity of graphene (Ganesan and Shaijumon 2015; Sham and Notley 2018).

The electrostatic interaction between the oxygen functionalities in rGO and the pollutants, especially various dyes, is the mechanism behind the adsorption. In the dye adsorption process, the vacancy sites and π–π interaction of conjugated structures of oxygen containing groups in the graphitic plans electrostatically interacted with the dyes in water (Liu et al. 2012; Yang et al. 2018; Zhang et al. 2016). The negatively charged surface is generally showing higher adsorption toward the cationic dye rather than the anionic dye molecules. The combination of non-covalent and electrostatic interactions is found to be responsible for the adsorption process of dyes on rGO (C.R. et al. 2017; Aboelfetoh, Elhelaly, and Gemeay 2018). The studies reported that the rGO synthesized by chemical method is better in adsorbing cationic dyes like methylene blue, methyl green (Figure 4.7), and anionic such as methyl orange dye. Many studies reveal that modified GO adsorbents are showing an enhanced adsorption capacity than that of unmodified form which obviously can attribute to an increase in the surface functional groups introduced by modification.

4.3.3   LDH for Dye Removal

LDH is a 2D anionic clay and is also referred to as hydrotalcite and because of the availability of a large range of composition/preparation variables, these clays have highly tunable brucite structures which are stacked in layers (Zubair et al. 2017). LDHs have a general formula of $[M_{1-x}{}^{2+}M_x{}^{3+}(OH)_2(A^{n-})_{x/n}]^{x+}\,mH_2O$, where M^{2+} and M^{3+} are divalent metal cation (e.g., Mg^{2+}, Co^{2+}, Ni^{2+}, Zn^{2+}, Cu^{2+}) and trivalent cations (e.g., Al^{3+}, Fe^{3+}, Ga^{3+}), respectively; x, varies from 0.20 to 0.33, and M^{2+}/M^{3+} molar ratio is 2.0–4.0. The excess positive charges are balanced by exchangeable anions in the interlayer galleries (e.g., $CO_3{}^{2-}$, Cl, NO_3, $SO_4{}^{2-}$) (Mittal 2021). It also contains water attached by a hydrogen bond to the hydroxide layer. It attained great interest due to its potential application in various files such as supercapacitors, catalysis, drug delivery, and water treatment, etc. This is due to their low cost, high surface area, highly tunable interlayer distance along with interior architecture, non-toxicity, and high anion exchange capacity (Yang et al. 2016). The LDH has a variety of compositions and structures which make it useful for the removal of various hazardous materials especially anionic materials. The economic viability and non-toxicity are also attractive features that make it widely used for removing anionic pollutants from water. Physical adsorption is the main mechanism in LDH and is take place by the intermolecular attraction between the adsorbent and adsorbate. Adsorption capacity and surface characteristics can be improved by modification with other materials such as carbon nanoparticles, metal particles, surfactants, polymers, etc. (Daud et al. 2019).

Through the adsorption process, the contaminants are adhered to the surface of the adsorbent. The surface forces or active sites on the surface of the adsorbent attract the adsorbate when it encounters the 2D material. The adsorption mechanism can be physical or chemical. In LDH, dye removal takes place mainly through the physical adsorption process and dyes are adsorbed on the surfaces of adsorbents because of the surface complexation and electrostatic interaction (Daud et al. 2019). The large surface area available for adsorption along with high anion exchange capacity and flexible interlayer galleries is the most favorable factor of LDH making it possible for adsorbing various ionic and no-ionic materials from contaminated solutions (Yang et al. 2016). There are different mechanisms involved in pollutant removals by LDH such as electrostatic attraction, ion exchange, van der Waals force, and hydrogen bonding interactions (Sansuk, Srijaranai, and Srijaranai 2016). LDH is found to be a good candidate for the removal of anionic dyes from water by exchanging the anions present in the interlayer galleries of the clay. Figure 4.8 shows the removal of acid yellow 25 by LDH. The adsorption capacities can be improved by modification of

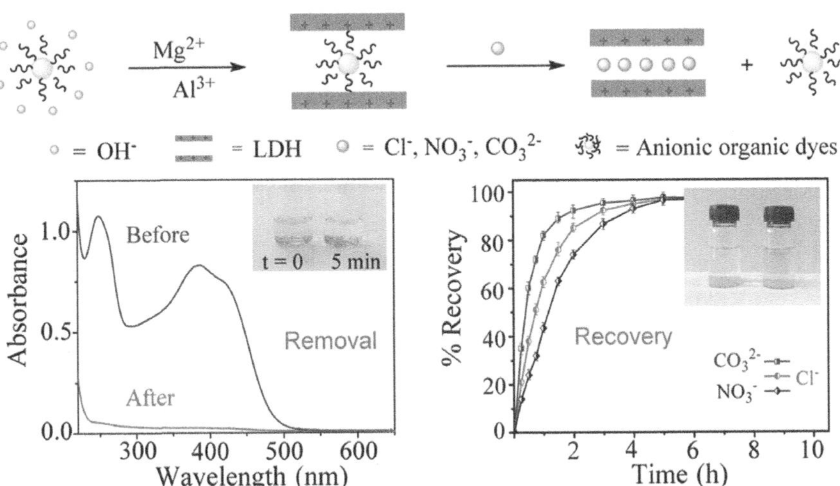

FIGURE 4.8 The removal of anionic organic dye by the electrostatic assembly and their recovery by the anionic exchange. Reproduced with permission from (Sansuk, Srijaranai, and Srijaranai 2016).

LDH by various methods and doping with metal ions is one of the good methods where the good distribution of metals in the hierarchical structure enhances the dye sorption inside the particle (Yang et al. 2016).

The adoption of LDH is also affected by various parameters such as pH, temperature, and various functional groups present in LDH (Yang et al. 2016). The surface morphology and performance of LDH are influenced by the pH and 3 or less than 3 pH favors the anionic dye removal because at this pH the adsorbent surface has an enhanced positive charge which facilitates a strong electrostatic interaction with anionic dyes thereby increasing the adsorption capacity.

The adsorption performance is also altered by the temperature because it influences the mobility and solubility of the dye (Zubair et al. 2017). Most of the reactions are endothermic and favor a high operating temperature because of the high mobility of the dye molecule. But for an exothermic process, the adsorption capacity and operating temperature are inversely proportional and by increasing temperature, the adsorption capacity decreases because the interactive forces between the dye species and the active site on the adsorbent surface reduce, consequently. The modification of LDH also increases the adsorption capacity due to an increase in surface area and functional group, which act as an anchor for the pollutant molecule, generated by modification.

4.4 CONCLUSIONS

2D-layered nanomaterials such as MoS_2, GO, MXenes, MOF, MMT clay, and LDHs are proven to be excellent sorbents for wastewater treatment by removing pollutants such as organic dye and HMIs. Physical and chemical interactions between the sorbents and pollutants are the key route toward pollutant decontamination. Physisorption involves van der Waal's interaction while chemisorption involves both electrostatic interaction and surface complexation with pollutants. The excellent adsorption properties can be attributed to the large effective surface area. Besides, the uptake adsorption capacity can be modulated by exploiting the surface functionalities of 2D materials such as sulfur-, oxygen-, and nitrogen-rich functionalities. The surface functionalities with lone pairs of electrons not only offer the active sites for pollutant interaction but also provide ease for functionalization with other functional materials which enhances the uptake capacity and rapid adsorption kinetics.

REFERENCES

Aboelfetoh, Eman F., Abeer A. Elhelaly, and Ali H. Gemeay. 2018. Synergistic effect of Cu(II) in the one-pot synthesis of reduced graphene oxide (rGO/CuxO) nanohybrids as adsorbents for cationic and anionic dyes. *Journal of Environmental Chemical Engineering* 6 (1):623–634.

Abu-Nada, Abdulrahman, Ahmed Abdala, and Gordon McKay. 2021. Removal of phenols and dyes from aqueous solutions using graphene and graphene composite adsorption: a review. *Journal of Environmental Chemical Engineering* 9 (5):105858.

Ahmad, Siti Zu Nurain, Wan Norharyati Wan Salleh, Ahmad Fauzi Ismail, Norhaniza Yusof, Mohd Zamri Mohd Yusop, and Farhana Aziz. 2020. Adsorptive removal of heavy metal ions using graphene-based nanomaterials: Toxicity, roles of functional groups and mechanisms. *Chemosphere* 248:126008.

Ai, Kelong, Changping Ruan, Mengxia Shen, and Lehui Lu. 2016. MoS$_2$ nanosheets with widened interlayer spacing for high-efficiency removal of mercury in aquatic systems. 26 (30):5542–5549.

Allen, Matthew J., Vincent C. Tung, and Richard B. Kaner. 2010. Honeycomb carbon: a review of graphene. *Chemical Reviews* 110 (1):132–145.

Awasthi, Anuradha, Pradip Jadhao, and Kanchan Kumari. 2019. Clay nano-adsorbent: structures, applications and mechanism for water treatment. *SN Applied Sciences* 1 (9):1076.

Belhouchat, N., H. Zaghouane-Boudiaf, and César Viseras. 2017. Removal of anionic and cationic dyes from aqueous solution with activated organo-bentonite/sodium alginate encapsulated beads. *Applied Clay Science* 135:9–15.

C.R., Minitha, Lalitha M, Jeyachandran Y.L, Senthilkumar L, and Rajendra Kumar R.T. 2017. Adsorption behaviour of reduced graphene oxide towards cationic and anionic dyes: co-action of electrostatic and $\pi - \pi$ interactions. *Materials Chemistry and Physics* 194:243–252.

Chen, Guodong, Meng Sun, Qin Wei, Yongfang Zhang, Baocun Zhu, and Bin Du. 2013. Ag$_3$PO$_4$/graphene-oxide composite with remarkably enhanced visible-light-driven photocatalytic activity toward dyes in water. *Journal of Hazardous Materials* 244–245:86–93.

Chen, Yajie, Xue Bai, and Zhengfang Ye. 2020. Recent progress in heavy metal ion decontamination based on metal–organic frameworks. *Nanomaterials (Basel)* 10 (8):1481.

Chen, Yucheng, Huachao Yang, Zhaojun Han, et al. 2022. MXene-based electrodes for supercapacitor energy storage. *Energy & Fuels* 36 (5):2390–2406.

Das, Rasel, Chad D. Vecitis, Agnes Schulze, et al. 2017. Recent advances in nanomaterials for water protection and monitoring. *Chemical Society Reviews* 46 (22):6946–7020.

Daud, Muhammad, Abdul Hai, Fawzi Banat, et al. 2019. A review on the recent advances, challenges and future aspect of layered double hydroxides (LDH) – containing hybrids as promising adsorbents for dyes removal. *Journal of Molecular Liquids* 288:110989.

Dong, Zhihui, Feng Zhang, Dong Wang, Xia Liu, and Jian Jin. 2015. Polydopamine-mediated surface-functionalization of graphene oxide for heavy metal ions removal. *Journal of Solid State Chemistry* 224:88–93.

Dutta, Soumi, Bramha Gupta, Suneel Kumar Srivastava, and Ashok Kumar Gupta. 2021. Recent advances on the removal of dyes from wastewater using various adsorbents: a critical review. *Materials Advances* 2 (14):4497–4531.

Elwakeel, Khalid Z., Ahmed M. Elgarahy, Ziya A. Khan, Muath S. Almughamisi, and Abdullah S. Al-Bogami. 2020. Perspectives regarding metal/mineral-incorporating materials for water purification: with special focus on Cr(vi) removal. *Materials Advances* 1 (6):1546–1574.

Fatima, Jawaria, Adnan Noor Shah, Muhammad Bilal Tahir, et al. 2022. Tunable 2D nanomaterials; their key roles and mechanisms in water purification and monitoring. *Frontiers in Environmental Science* 10: 1–23, Article 766743.

Fu, Liangjie, Zhaoli Yan, Qihang Zhao, and Huaming Yang. 2018. Novel 2D nanosheets with potential applications in heavy metal purification: a review. *Advanced Materials Interfaces* 5 (23):1801094.

Ganesan, Aswathi, and Manikoth Shaijumon. 2015. Activated graphene-derived porous carbon with exceptional gas adsorption properties. *Microporous and Mesoporous Materials* 220:21–27.

Guo, Ting, Chaoke Bulin, Zeyu Ma, et al. 2021. Mechanism of Cd(II) and Cu(II) adsorption onto few-layered magnetic graphene oxide as an efficient adsorbent. *ACS Omega* 6 (25):16535–16545.

Gupta, Deepika, Vishnu Chauhan, and Rajesh Kumar. 2020. A comprehensive review on synthesis and applications of molybdenum disulfide (MoS$_2$) material: past and recent developments. *Inorganic Chemistry Communications* 121:108200.

He, Hongping, Lingya Ma, Jianxi Zhu, Ray L. Frost, Benny K. G. Theng, and Faïza Bergaya. 2014. Synthesis of organoclays: a critical review and some unresolved issues. *Applied Clay Science* 100:22–28.

Huang, Peng, Algy Kazlauciunas, Robert Menzel, and Long Lin. 2017. Determining the mechanism and efficiency of industrial dye adsorption through facile structural control of organo-montmorillonite adsorbents. *ACS Applied Materials & Interfaces* 9 (31):26383–26391.

Isfahani, Ali Pournaghshband, Ahmad A. Shamsabadi, Farbod Alimohammadi, and Masoud Soroush. 2022. Efficient mercury removal from aqueous solutions using carboxylated Ti3C2Tx MXene. *Journal of Hazardous Materials* 434:128780.

Ishag, Alhadi, and Yubing Sun. 2021. Recent advances in two-dimensional MoS_2 nanosheets for environmental application. *Industrial & Engineering Chemistry Research* 60 (22):8007–8026.

Jeon, Minjung, Byung-Moon Jun, Sewoon Kim, et al. 2020. A review on MXene-based nanomaterials as adsorbents in aqueous solution. *Chemosphere* 261:127781.

Jia, Feifei, Qingmiao Wang, Jishan Wu, Yanmei Li, and Shaoxian Song. 2017. Two-dimensional molybdenum disulfide as a superb adsorbent for removing Hg^{2+} from water. *ACS Sustainable Chemistry & Engineering* 5 (8):7410–7419.

Kausar, Abida, Munawar Iqbal, Anum Javed, et al. 2018. Dyes adsorption using clay and modified clay: a review. *Journal of Molecular Liquids* 256:395–407.

Kumar, Neeraj, Elvis Fosso-Kankeu, and Suprakas Sinha Ray. 2019. Achieving controllable MoS_2 nanostructures with increased interlayer spacing for efficient removal of Pb(II) from aquatic systems. *ACS Applied Materials & Interfaces* 11 (21):19141–19155.

Li, Guiliang, Yang Liu, Yi Shen, Qile Fang, and Fu Liu. 2021. Bimetallic coordination in two-dimensional metal–organic framework nanosheets enables highly efficient removal of heavy metal lead (II). *Frontiers in Chemical Engineering* 3:636439.

Li, Jie, Qingyun Duan, Zheng Wu, et al. 2020. Few-layered metal-organic framework nanosheets as a highly selective and efficient scavenger for heavy metal pollution treatment. *Chemical Engineering Journal* 383:123189.

Li, Zeyang, Ruoyu Fan, Zhi Hu, et al. 2020. Ethanol introduced synthesis of ultrastable 1T-MoS_2 for removal of Cr(VI). *Journal of Hazardous Materials* 394:122525.

Liu, Fei, Soyi Chung, Gahee Oh, and Tae Seok Seo. 2012. Three-dimensional graphene oxide nanostructure for fast and efficient water-soluble dye removal. *ACS Applied Materials & Interfaces* 4 (2):922–927.

Liu, Jiadi, Qingqing Li, Feifei Mao, Kuaibing Wang, and Hua Wu. 2021. 2D MOFs-based materials for the application of water pollutants removing: fundamentals and prospects. *Chemistry: An Asian Journal* 16 (22):3585–3598.

Liu, Zhihang, Qian Wang, Xiujie Huang, and Xueren Qian. 2022. Surface functionalization of graphene oxide with hyperbranched polyamide-amine and microcrystalline cellulose for efficient adsorption of heavy metal ions. *ACS Omega* 7 (13):10944–10954.

Luo, Jinming, Kaixing Fu, Meng Sun, et al. 2019. Phase-mediated heavy metal adsorption from aqueous solutions using two-dimensional layered MoS_2. *ACS Applied Materials & Interfaces* 11 (42):38789–38797.

Mittal, J. 2021. Recent progress in the synthesis of layered double hydroxides and their application for the adsorptive removal of dyes: a review. *J Environ Manage* 295:113017.

Muhammad Ekramul Mahmud, Habibun Nabi, A. K. Obidul Huq, and Rosiyah binti Yahya. 2016. The removal of heavy metal ions from wastewater/aqueous solution using polypyrrole-based adsorbents: a review. *RSC Advances* 6 (18):14778–14791.

Naik, Smita Gajanan, and Mohammad Hussain K. Rabinal. 2020. Molybdenum disulphide heterointerfaces as potential materials for solar cells, energy storage, and hydrogen evolution. *Energy Technology* 8 (6):1901299.

Ngulube, T., J. R. Gumbo, V. Masindi, and A. Maity. 2017. An update on synthetic dyes adsorption onto clay based minerals: a state-of-art review. *Journal of Environmental Management* 191:35–57.

Nováček, Michal, Ondřej Jankovský, Jan Luxa, et al. 2017. Tuning of graphene oxide composition by multiple oxidations for carbon dioxide storage and capture of toxic metals. *Journal of Materials Chemistry A* 5 (6):2739–2748.

Othman, Zakarya, Hamish R. Mackey, and Khaled A. Mahmoud. 2022. A critical overview of MXenes adsorption behavior toward heavy metals. *Chemosphere* 295:133849.

Peng, Na, Danning Hu, Jian Zeng, Yu Li, Lei Liang, and Chunyu Chang. 2016. Superabsorbent cellulose–clay nanocomposite hydrogels for highly efficient removal of dye in water. *ACS Sustainable Chemistry & Engineering* 4 (12):7217–7224.

Peng, Qiuming, Jianxin Guo, Qingrui Zhang, et al. 2014. Unique lead adsorption behavior of activated hydroxyl group in two-dimensional titanium carbide. *Journal of the American Chemical Society* 136 (11):4113–4116.

Peng, Weijun, Hongqiang Li, Yanyan Liu, and Shaoxian Song. 2017. A review on heavy metal ions adsorption from water by graphene oxide and its composites. *Journal of Molecular Liquids* 230:496–504.

Perreault, François, Andreia Fonseca de Faria, and Menachem Elimelech. 2015. Environmental applications of graphene-based nanomaterials. *Chemical Society Reviews* 44 (16):5861–5896.

Rashid, Ruhma, Iqrash Shafiq, Parveen Akhter, Muhammad Javid Iqbal, and Murid Hussain. 2021. A state-of-the-art review on wastewater treatment techniques: the effectiveness of adsorption method. *Environmental Science and Pollution Research* 28 (8):9050–9066.

Saleh, Muhammad, Vimlesh Chandra, K. Christian Kemp, and Kwang S. Kim. 2013. Synthesis of N-doped microporous carbon via chemical activation of polyindole-modified graphene oxide sheets for selective carbon dioxide adsorption. *Nanotechnology* 24 (25):255702.

Sansuk, Sira, Somkiat Srijaranai, and Supalax Srijaranai. 2016. A new approach for removing anionic organic dyes from wastewater based on electrostatically driven assembly. *Environmental Science & Technology* 50 (12):6477–6484.

Shahzad, Asif, Kashif Rasool, Waheed Miran, et al. 2017. Two-dimensional Ti_3C_2Tx MXene nanosheets for efficient copper removal from water. *ACS Sustainable Chemistry & Engineering* 5 (12):11481–11488.

Sham, Alison Y. W., and Shannon M. Notley. 2018. Adsorption of organic dyes from aqueous solutions using surfactant exfoliated graphene. *Journal of Environmental Chemical Engineering* 6 (1):495–504.

Sharma, Ponchami, and Manash R. Das. 2013. Removal of a cationic dye from aqueous solution using graphene oxide nanosheets: investigation of adsorption parameters. *Journal of Chemical & Engineering Data* 58 (1):151–158.

Sharma, Prerana, Anuj K. Singh, and Vinod K. Shahi. 2019. Selective adsorption of Pb(II) from aqueous medium by cross-linked chitosan-functionalized graphene oxide adsorbent. *ACS Sustainable Chemistry & Engineering* 7 (1):1427–1436.

Sharma, S., and A. Bhattacharya. 2017. Drinking water contamination and treatment techniques. *Applied Water Science* 7 (3):1043–1067.

Sheth, Yashvi, Swapnil Dharaskar, Vishal Chaudhary, Mohammad Khalid, and Rashmi Walvekar. 2022. Prospects of titanium carbide-based MXene in heavy metal ion and radionuclide adsorption for wastewater remediation: a review. *Chemosphere* 293:133563.

Sophia A, Carmalin, and Eder C. Lima. 2018. Removal of emerging contaminants from the environment by adsorption. *Ecotoxicology and Environmental Safety* 150:1–17.

Sun, Huating, Tianxing Wu, Yunxia Zhang, Dickon H. L. Ng, and Guozhong Wang. 2018. Structure-enhanced removal of Cr(vi) in aqueous solutions using MoS_2 ultrathin nanosheets. *New Journal of Chemistry* 42 (11):9006–9015.

Unuabonah, Emmanuel I., Christina Günter, Jens Weber, Susanne Lubahn, and Andreas Taubert. 2013. Hybrid clay: a new highly efficient adsorbent for water treatment. *ACS Sustainable Chemistry & Engineering* 1 (8):966–973.

Velusamy, Sasireka, Anurag Roy, Senthilarasu Sundaram, and Tapas Kumar Mallick. 2021. A review on heavy metal ions and containing dyes removal through graphene oxide-based adsorption strategies for textile wastewater treatment. *The Chemical Record* 21 (7):1570–1610.

Wang, Qi, and Didier Astruc. 2020. State of the art and prospects in metal–organic framework (MOF)-based and MOF-derived nanocatalysis. *Chemical Reviews* 120 (2):1438–1511.

Xie, Xiuqiang, Chi Chen, Nan Zhang, Zi-Rong Tang, Jianjun Jiang, and Yi-Jun Xu. 2019. Microstructure and surface control of MXene films for water purification. *Nature Sustainability* 2 (9):856–862.

Xu, Rongming, Meipeng Jian, Qinghua Ji, et al. 2020. 2D water-stable zinc-benzimidazole framework nanosheets for ultrafast and selective removal of heavy metals. *Chemical Engineering Journal* 382:122658.

Yang, Kaijie, Jun Wang, Xiaoxiao Chen, Qiang Zhao, Abdul Ghaffar, and Baoliang Chen. 2018. Application of graphene-based materials in water purification: from the nanoscale to specific devices. *Environmental Science: Nano* 5 (6):1264–1297.

Yang, Zhongzhu, Fenghua Wang, Chang Zhang, et al. 2016. Utilization of LDH-based materials as potential adsorbents and photocatalysts for the decontamination of dyes wastewater: a review. *RSC Advances* 6 (83):79415–79436.

Ye, Shibing, Yue Liu, and Jiachun Feng. 2017. Low-density, mechanical compressible, water-induced self-recoverable graphene aerogels for water treatment. *ACS Applied Materials & Interfaces* 9 (27):22456–22464.

Ying, Yulong, Yu Liu, Xinyu Wang, et al. 2015. Two-dimensional titanium carbide for efficiently reductive removal of highly toxic chromium(VI) from water. *ACS Applied Materials & Interfaces* 7 (3):1795–1803.

Zhang, Yanyang, Bing Wu, Hui Xu, et al. 2016. Nanomaterials-enabled water and wastewater treatment. *NanoImpact* 3–4:22–39.

Zhao, Jian, Zhenyu Wang, Jason C. White, and Baoshan Xing. 2014. Graphene in the aquatic environment: adsorption, dispersion, toxicity and transformation. *Environmental Science & Technology* 48 (17):9995–10009.

Zhu, Runliang, Qingze Chen, Qing Zhou, Yunfei Xi, Jianxi Zhu, and Hongping He. 2016. Adsorbents based on montmorillonite for contaminant removal from water: a review. *Applied Clay Science* 123:239–258.

Zubair, Mukarram, Muhammad Daud, Gordon McKay, Farrukh Shehzad, and Mamdouh A. Al-Harthi. 2017. Recent progress in layered double hydroxides (LDH)-containing hybrids as adsorbents for water remediation. *Applied Clay Science* 143:279–292.

5 Engineering 2D Semiconductors as Promising Materials for Solar Fuel Generation by CO$_2$ Photoreduction

V. P. Animasree, Vishal Kandathil
and Narayanapillai Manoj

5.1 INTRODUCTION

Global warming has become a nightmare for the whole living system in recent times, with the reason behind being human activities like rapid industrialization and urbanization (Khan et al. 2020; Dong et al. 2019). Climate change and its effects all over the globe, which can be seen every day now, are mainly because of global warming (Aizebeokhai 2009). The effect of climate change is greater than expected and is a direct threat to food security. It also results in rising sea levels, floods, and droughts and causes severe damage to economic conditions (Mishra, Singh, and Jain 2010; McCarthy, Best, and Betts 2010). The greenhouse gases contribute much to global warming, with carbon dioxide (CO$_2$) being the main among them. The natural and sustainable way of maintaining the carbon cycle in nature is by capturing it and converting it to carbohydrates by plants through the process of photosynthesis. But since urbanization has led to large-scale deforestation thereby increasing the levels of atmospheric CO$_2$ and also since there is a many-fold increase in the release of anthropogenic CO$_2$ into the atmosphere, the trees may not fully capture those (López-Carr and Burgdorfer 2013; Di Sacco et al. 2021). Hence, it is urgent that new strategies for regulating atmospheric CO$_2$ levels be developed and put into place.

Governments and organizations are taking a variety of measures to regulate the amounts of CO$_2$ and other greenhouse gases present in the atmosphere. Aside from measures such as increasing the use of renewable energy, lowering carbon emissions by improving vehicle fuel efficiency, planting more trees, and so on, the scientific community is already conducting extensive research to control CO$_2$ levels by capturing and converting it into useful products (Dincer 2000; Güney 2019). The process for CO$_2$ utilization involving catalysts can be thermochemical, electrochemical, or photochemical pathways (Roy, Cherevotan, and Peter 2018; Whipple and Kenis 2010; Yaashikaa

DOI: 10.1201/9781003343899-5

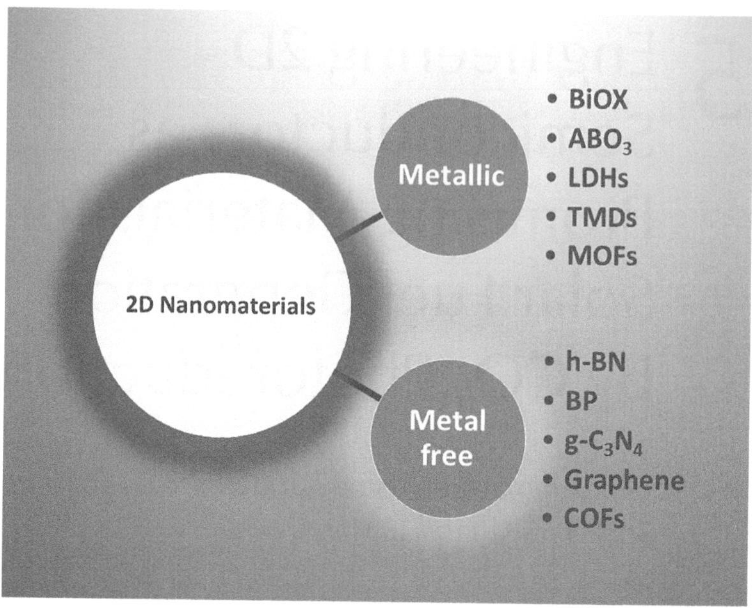

FIGURE 5.1 Metallic and metal-free 2D nanomaterials for various applications.

et al. 2019; Hu, Guild, and Suib 2013). For the utilization of CO_2, catalysts can play an important role. When developing a catalyst for CO_2 capture and utilization, researchers need to pay close attention to a wide range of aspects, including the catalyst's thermal stability, porosity, pH, surface area, and cost, amongst other considerations.

Recent research indicates that two-dimensional (2D) nanomaterials are indeed an ideal choice as catalysts and supports for a variety of organic transformations and CO_2 utilization (Rosso et al. 2021; Tan et al. 2017). As shown in Figure 5.1, there are metallic and metal-free 2D nanomaterials that can be investigated in the field of catalysis. The unique structural and electronic properties of 2D nanomaterials make them outstanding materials for catalysis (Rosso et al. 2021). The improved intrinsic features of 2D nanomaterials can be due to the quantum confinement effect and surface effect, along with anisotropy (Angizi et al. 2022; Jelmy et al. 2021). However, there is a lot of room for improvement in the photocatalytic behavior of pristine 2D nanomaterials, which can create hindrances to their practical applications. Surface defect engineering can be implemented to improve the photocatalytic efficiencies of 2D nanomaterials. When compared to their bulk counterparts, ultrathin 2D nanomaterials with large surface areas can produce more charge carriers by effective light harvesting (Tan et al. 2017). The architecture of ultrathin 2D nanomaterials reduces the likelihood of electron (e^-)-hole (h^+) pair recombination, which aids in reducing the charge diffusion distance between the surface and the bulk.

5.2 PHOTOCATALYSIS

Photocatalysts are materials that can change the rate at which chemicals react in the presence of light, and the process involved is called photocatalysis. Photocatalysis involves the utilization of light and a semiconductor, i.e., the photocatalyst, for the reaction to occur (Mills and Le Hunte 1997). Upon exposure of the semiconductor material to light, an e^--h^+ pair is generated, which is how the reaction starts. Photocatalysis, like conventional catalysis, can be classified into two types: homogeneous and heterogeneous, depending on their physical state in the reaction medium.

5.2.1 Homogeneous Photocatalysis

Homogeneous photocatalysis is a type of photocatalysis in which the photocatalyst and the reactants are present in the same phase during the catalysis process. Homogeneous photocatalysts have many benefits, including low cost and high functionality under ambient conditions, low optical scattering, excellent light absorption, and high photo-transport phenomena (Cieśla et al. 2004). Apart from the applications including photocatalytic depollution, homogeneous photocatalysts find applications in the photoreduction of CO_2 and photocatalytic water splitting. These photocatalysts are capable of performing both the role of solar harvesters and the role of electronically excited molecules to initiate a chemical reaction through the process of photoinduced electron transfer (Dadashi-Silab, Doran, and Yagci 2016). The photocatalysts help in the conversion of chemical bonds in the reactants to convert them into the required products. Even though these homogeneous photocatalysts are extremely active and effective, their use on a commercial scale is hampered by several drawbacks, including the tedious workup and recovery of the catalyst, the photostability of the catalyst, and product poisoning caused by these catalysts.

5.2.2 Heterogeneous Photocatalysis

Heterogeneous photocatalysis refers to a catalytic process in which the photocatalyst and the reactants exist in physically distinct phases. The catalytically active centers are usually dispersed in heterogeneous support or the support itself acts as an active center in the case of heterogeneous photocatalysis (Fox and Dulay 1993). The heterogeneous surface of the catalyst makes available an environment that influences the chemical reactivity of a wide range of reactants or adsorbents and helps in initiating photoinduced redox reactivity in molecules. Generally, the semiconductor catalyst gets activated by absorbing light in a wavelength region that is different from that of the substrate of interest. Therefore, photosensitization occurs in the process of heterogeneous photocatalysis, which involves the indirect photoactivation of the heterogeneous catalyst rather than the direct formation of the excited state of the substrate. The inertness toward chemical environments and the photostability along with the ease of recycling makes the heterogeneous photocatalysis more attractive and can be scaled up even though there is a drop in the efficiency of activity when compared with its homogeneous counterparts. Heterogeneous photocatalysis finds many practical applications, ranging from pharmaceuticals to food industries, sunscreens to solar cells. Titanium dioxide (TiO_2) is the most extensively studied heterogeneous photocatalyst with a wide range of applications (Padmanabhan et al. 2021; Padmanabhan, Gopalakrishnan, and John 2022).

The materials are classified mainly into three classes depending upon the bandgap (E_g), i.e., the energy difference between the valence band and the conduction band (Figure 5.2). They are as follows: metal or conductor (E_g value of $<1.0\,eV$), semiconductor (E_g value of <1.5–$3.0\,eV$), and insulator (E_g value of $>5.0\,eV$) (Ameta et al. 2018). Since they can conduct electricity at even room temperature in the presence of light, semiconductors can be used as photocatalysts among these materials. The electrons in the valence band get excited to the conduction band by the absorption of energy from photons upon exposure to light of the desired wavelength by creating holes in the valence band. As a result, an electron-hole pair is generated, and the process of photocatalysis ensues.

Depending on the redox levels and the positions of the valence band and conduction band, the semiconductor and the substrate can interact in different ways. The substrate may undergo reduction if its redox level is less than that of the conduction band of the catalyst, whereas the substrate will undergo oxidation if its redox level is greater than that of the valence band. Whenever the redox level of the material is greater than that of the conduction band of the catalyst but less than the valence band, neither oxidation nor reduction is possible, and vice versa.

This chapter aims to give a brief overview of the recent advances in the field of CO_2 photoreduction using 2D nanomaterials. Also, the key factors affecting the photocatalytic activity of the 2D nanomaterials like surface area, pore structure and volume, morphology, and so on are looked into. This chapter also sheds light on the photocatalytic reduction mechanism. Finally, a summary of the literature is provided, along with some perspectives for future work.

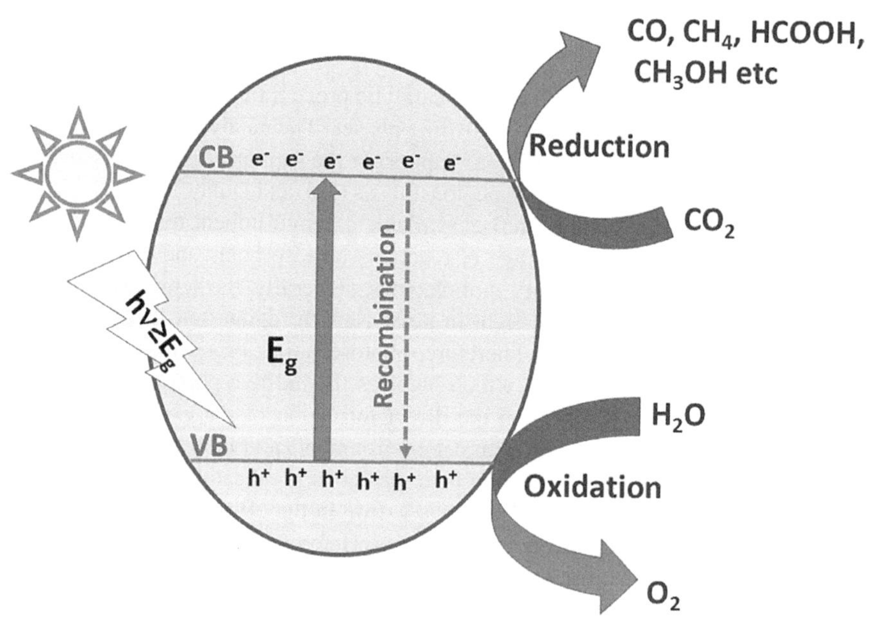

FIGURE 5.2　Representation of energy bands.

5.3 PHOTOCATALYTIC REDUCTION OF CO_2

Photocatalytic reduction is the most effective method for lowering CO_2 levels in the atmosphere (Chen and Jin 2019). Semiconductor photocatalysis is an emerging area that includes various semiconductor photocatalysts like TiO_2, graphitic carbon nitride (g-C_3N_4), graphene oxide (GO), molybdenum disulfide (MoS_2), etc. Photocatalysis is considered to be an alternative to photosynthesis (Ali, Razzaq, and In 2019; Sato et al. 2021). In the presence of sunlight, photocatalysts are employed to achieve the primary goal of converting CO_2 into value-added chemicals and solar fuels such as carbon monoxide (CO), methane (CH_4), formic acid (HCOOH), formaldehyde (HCHO), and methanol (CH_3OH) (Yuan and Xu 2015; Yang et al. 2021; Ali, Razzaq, and In 2019; Koči et al. 2020). Photocatalysis has a significant advantage when compared with other methods because it can occur under ambient conditions (Al Jitan, Palmisano, and Garlisi 2020).

5.3.1 General Mechanism of CO_2 Photoreduction

CO_2 is thermodynamically the most stable linear molecule, with a dissociation energy of 750 kJ/mol (Xiong et al. 2019). It is much more difficult to break the C=O bond than other bonds like C–C ($\sim$336 kJ/mol), C–O ($\sim$327 kJ/mol), and C–H ($\sim$411 kJ/mol) (Lin et al. 2020). It shows the highest oxidation state of the carbon atom with a Gibbs free energy of $\otimes G^\circ = -394.4$ kJ/mol. A more negative potential is necessary to break the C=O bond and bend the linear geometry of CO_2, which accelerates the spontaneous transformation of CO_2 (Nguyen et al. 2020). Various products obtained from the CO_2 reduction and corresponding thermodynamic potentials are given in Table 5.1, at pH 7 in an aqueous solution versus a normal hydrogen electrode (NHE), at 25 °C, and 1 atm gas pressure (Kumar et al. 2012; Lin et al. 2020; Nguyen et al. 2020).

More intriguing factors exist in heterogeneous photocatalysis employing semiconductors. A bandgap level positioned at the top of the filled valence band and the bottom of the unoccupied conduction band is more appropriate for performing a CO_2 photoreduction reaction than other materials (Mao, Li, and Peng 2013). In general, there are three main steps for the semiconductor photocatalytic reduction of CO_2. First, the light energy (hυ) greater than or equal to the E_g of the semiconductor photocatalyst it

TABLE 5.1

Products obtained from CO_2 reduction with corresponding thermodynamic potentials

Reaction	Product	E° (V vs. NHE) (V)
$CO_2 + e^- \rightarrow CO_2{\cdot}^-$	Carbon dioxide radical anion	-1.90
$CO_2 + 2H^+ + 2e^- \rightarrow HCOOH$	Formic acid	-0.61
$CO_2 + 2H^+ + 2e^- \rightarrow CO + H_2O$	Carbon monoxide	-0.53
$CO_2 + 4H^+ + 4e^- \rightarrow HCHO + H_2O$	Formaldehyde	-0.48
$CO_2 + 6H^+ + 6e^- \rightarrow CH_3OH + H_2O$	Methanol	-0.38
$CO_2 + 8H^+ + 8e^- \rightarrow CH_4 + 2H_2O$	Methane	-0.24

absorbs from the sun is used to produce e^--h^+ pairs. The second is charge separation. During charge separation, an e^- in the valence band of the semiconductor photocatalyst migrates to its conduction band and leaves h^+ in the valence band. The e^--h^+ pair are charge carriers and are collectively called excitons. Finally, the charge carriers are transported to the surface of the semiconductor photocatalysts where they undergo redox reactions (Figure 5.3) (Lingampalli, Ayyub, and Rao 2017).

The reaction rate of the photocatalytic process greatly depends on factors such as the nature of the photocatalyst, the energy of light, pH, and reactant concentration. Once CO_2 is adsorbed on the surface of the photocatalyst, its structure changes to an active bent form from a linear form, followed by the binding of carbon atoms from the CO_2 molecule to the oxygen sites on the surface, as well as the binding of oxygen atoms from the CO_2 molecule to the metal center of the catalyst. During the structural transformation, the activation energy decreases due to a decrease in the metal oxide's lowest unoccupied molecular orbital level (Al Jitan, Palmisano, and Garlisi 2020). More specifically, after the light illumination, the generated photoexcited electrons react with the CO_2 adsorbed on the surface of the photocatalyst and the protons supplied by the reducing agents to form the desired products (Nguyen et al. 2020).

Multiple steps are involved in selecting photocatalysts, all of which are intended to simplify the process of photoreduction. The majority of metal oxide semiconductor catalysts are under a wide bandgap range. So, the absorption is from the ultraviolet (UV) region. This limits the use of the visible solar spectrum. The fast recombination process results in the decreasing efficiency of the photocatalyst. This is the major drawback of semiconductor photocatalysts. The following are the requirements for a good photocatalyst to undergo a photoreduction reaction: (i) more positive valance band to be acting as an electron acceptor, (ii) sufficiently negative conduction band with adequate electrons density, concerning the onset reduction potential of CO_2, (iii) the high adsorption capability of reactants (CO_2 or CO_3^{2-}) onto the photocatalyst (iv) scarcity in toxic photochemical reaction mediated byproduct formation, and (v) adequate stability of the developed photocatalyst (Shit et al. 2020).

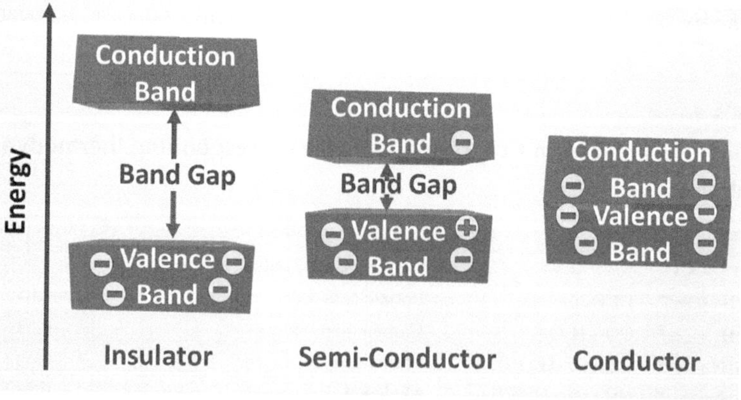

FIGURE 5.3 Schematic representation of the photocatalytic CO_2 reduction mechanism.

Doping is the most widely accepted tool to reduce the e^--h^+ pair recombination and to enhance the photocatalytic reduction reaction. Broadly used dopants are categorized into metals (copper, nickel, gold, silver, cerium, tin, etc.) and non-metals (nitrogen, sulfur, etc.) (Al Jitan, Palmisano, and Garlisi 2020). Doping with metals and non-metals primarily produces an inter-band state and modifies the band structures that allow for charge transfer with adsorbed CO_2 molecules (Shit et al. 2020).

5.4 2D NANOMATERIALS FOR PHOTOCATALYSIS

Photocatalysis is the most effective method for reducing atmospheric CO_2 because light energy is the primary source. Surface engineering of 2D nanomaterials allows for increased efficiency and product selectivity in the CO_2 photoreduction reaction. The effective generation of excitons, separation of photogenerated charge, CO_2 adsorption, and desorption capabilities can be enhanced by the surface or structural modifications of the 2D nanomaterials. Defect engineering, which is considered to be an effective surface engineering strategy, can be achieved primarily by plasma or chemical treatments (Hu et al. 2018; Jiang et al. 2019). Plasma consists of a system composed of charged and movable particles. These kinetically energetic charged active ions can interact with the materials and modify their structure by introducing vacancies and structural distortion. Also, plasma can interact at the defect sites of the 2D nanomaterials, forming adatoms or impurities. The plasma irradiation can be controlled by controlling factors like pressure, time, and power (Yang et al. 2014; Amani et al. 2015; Sun et al. 2018). Chemical treatment can also induce defects in 2D nanomaterials without causing significant material damage. By selectively choosing the reactants, chemical treatment can improve charge transfer, stimulate efficient doping, and alter the structure's band and phase. In addition to plasma and chemical treatment approaches, laser modification, as well as ozone reactions, are also utilized in defect engineering. This section highlights the significance of various 2D nanomaterials functioning as photocatalysts. Research in this field had already begun many years ago. Xiaoyu Chen and colleagues used the solid-liquid phase arc discharge technique to create ultrathin, single-crystal tungsten trioxide (WO_3) nanosheets with a thickness of 4–5 nm (Chen et al. 2012).

The graphene-based nanocomposites show an improved photocatalytic performance which can be attributed to several factors, including the lowering of e^--h^+ pair recombination, the expansion of the light absorption spectrum, the increase in light intensity, the enhancement of surface-active sites, and the improvement of the chemical stability of photocatalysts. Graphene-based photocatalysts have wide applicability, including water splitting, the oxidation or degradation of organic pollutants, the photoconversion of CO_2 into sustainable fuels, the removal of harmful heavy metal ions, and antibacterial applications (Tu, Zhou, and Zou 2013). Due to its long-term stability against photo and chemical corrosion, low cost, non-toxicity, and powerful oxidizing and reducing power, TiO_2 has received extensive attraction as a semiconductor photocatalyst (Yui et al. 2011). Table 5.2 shows the fabrication methods and reaction conditions employed by different photocatalysts. The photoreduction of CO_2 was significantly aided by the combination of graphene and TiO_2. Xiaoyang Pan and co-workers developed free-stranded TiO_2 nanosheets using a graphene template by a bottom-up strategy, and benzyl alcohol was used to functionalize the graphene. Precursors such as tetrabutyl-orthotitanate (TBOT) and cadmium nitrate ($Cd(NO_3)_2$) were used for this study. Sodium borohydride ($NaBH_4$) is utilized as the reducing agent for the conversion of GO to graphene via the ultrasonication method, which was accomplished by a modified Hummer's process and graphite powder as the starting material (Zaaba et al. 2017). The synergistic interaction of graphene with benzyl alcohol should lead to the production of TiO_2 2D nanosheets. In other words, graphene acts as a 2D scaffold and benzyl alcohol produces a uniform coating of TiO_2 on the graphene nanosheets. The graphene nanosheet template was eliminated following heat treatment in the air, and then TiO_2 2D nanosheets were produced. Since TiO_2 has a large bandgap and cadmium sulfide (CdS) has a low bandgap, their connectivity increases the capacity for light absorption and the interactions between e^--h^+ pairs for surface reactions. Therefore, CdS was efficiently deposited as a result of the photodeposition process, forming CdS-TiO_2 nanosheets, which enhances the photocatalytic effectiveness. In this study, scientists compared TiO_2 nanosheets and CdS-TiO_2 nanosheets. They found that CdS-TiO_2 nanosheets were better at photocatalysis. Photocatalytic CO_2 conversion was done in a stainless-steel reactor with quartz glass on top. Water, which can act as an electron donor, was used, and a 300W Xe arc lamp was used as the light source. A gas (vapor)-solid heterogeneous reaction mechanism was used to complete the CO_2 photoreduction. Gas chromatography (GC) with a flame ionization detector (FID) was used to estimate the quantity of CO and CH_4 generated. A methanation reactor further converted carbon monoxide (CO) and CO_2 to CH_4, which was then measured by the FID. When bare TiO_2 nanosheet performance was compared to CdS-modified TiO_2 nanosheet performance, the conversion of CO_2 to CO and CH_4 is higher in the case of the CdS-TiO_2 nanosheet, with values of 3.64 and 0.54 µmol/g/h, respectively (Pan and Xu 2015).

Reduced graphene oxide (rGO) is a class of functional materials that is also trending in the conversion and storage of light energy. Composites made of semiconductors and rGO can increase the effectiveness of photocatalysis. By adopting a surface modification approach and 4-aminothiophenol, Rajesh Bera et al. effectively produced various dimensions of CdS, which were grafted onto the rGO and studied the effect of photocatalytic properties. The GO was made by using a modified Hummer's method; cadmium chloride was employed as the source of cadmium in the process, and CdS nanosheets were prepared by the hydrothermal method. The transmission electron microscopy data shows the different dimensions of the CdS nanostructure with rGO, and the surface modification with 4-aminothiophenol was examined using Fourier transform infrared spectroscopy (FT-IR) and confirmed by the X-ray photoelectron spectroscopy (XPS) technique. According to the zeta potential data, the positive potential value of various samples with CdS surfaces makes it easier for electrostatic interactions with negatively charged GO (−25 mV) surfaces to occur, leading to CdS-GO composites. The results of a study on photo dye degradation indicate that maximum degradation occurs at nanosheet samples, followed by nanorod and nanoparticle samples and that efficiency also occurs in the same sequence (Bera, Kundu, and Patra 2015).

Recently, g-C_3N_4 has shown very important aspects of CO_2 photoreduction. Low cost, easy preparation method, high thermal and chemical stability promote the activity of photoreduction. The conduction band potential of g-C_3N_4 is $-1.20\,V$ (versus normal hydrogen electrode, NHE at pH $= 7$), which effectively reduces CO_2 into valuable hydrocarbon fuels like HCOOH with a standard redox potential of $(0.61\,V)$, HCHO $(-0.48\,V)$, CH_3OH $(-0.38\,V)$, and CH_4 $(-0.24\,V)$. The photocatalytic efficiency of g-C_3N_4 seems to be too low because of the low charge transfer efficiency and high recombination rate of the photoexcited e^--h^+ pairs. To overcome this issue, there are other prominent methods widely used, such as coupling of g-C_3N_4 with another semiconductor, usage of photosensitizers, or co-catalysts. However, the development of a direct Z-scheme system enhances the power of g-C_3N_4. Thanh Truc and co-workers studied the role of niobium (Nb) in the TiO_2 lattice, which reduces the bandgap energy or enhances the photocatalytic activity of the Nb-TiO_2 (Truc et al. 2019). The prepared Nb-TiO_2 was then employed to hybridize with g-C_3N_4 to develop visible light active Nb-TiO_2/g-C_3N_4 direct Z-scheme system which facilitated the superior reduction of CO_2 into valuable fuels. Here, the CB of the g-C_3N_4 has a potential energy of $-1.2\,V$ which contains an e^- of the Nb-TiO_2/g-C_3N_4. The e^- was strong enough for the reduction of CO_2 to generate not only CH_4 and CO but also HCOOH.

For the study of the conversion of CO_2 into methane under visible light irradiation, Wee-Jun Ong and co-workers created a sandwich model graphene/g-C_3N_4 nanohybrid (GCN) catalyst using a facile one-pot impregnation-thermal reduction method. Direct polymerization of urea was employed for the successful production of the catalyst structure, and GO was used as the structure-directing agent. The 8-electron reaction that converts CO_2 to CH_4 was made possible by the involvement of graphene, which acts as a suitable frame and lowers the rate at which e^--h^+ pairs recombine, accelerating the formation of the desired product. In comparison to bare g-C_3N_4 material, the graphene-incorporated material had increased activity in terms of CH_4 production under visible light irradiation. Several samples were prepared based on the weight percentage of graphene, with the sample containing 15 wt.% graphene (GCN 0.15) showing the highest conversion at 5.87 µmol/g at ambient conditions (Ong et al. 2015).

By applying the wet chemical impregnation method, Lling-Lling Tan and co-workers synthesized a GO-impregnated TiO_2 sample named GO-OTiO_2 and studied its activity in the synthesis of CH_4 by the photoreduction of CO_2 (Figure 5.4). This sample's unique feature is its high oxygen concentration, which aids in the conversion process more effectively. The research group initially created oxygen-rich TiO_2 using a standard aqueous peroxo-titanate method, with titanium butoxide serving as the titanium precursor. The prepared sample displayed features that were active under visible light. However, the photoactivity investigation shows that the bare O_2-TiO_2 sample degenerates after a prolonged response. To solve this issue, GO was added to the sample and owing to its special features, the performance was improved. The GO-OTiO_2 sample yielded 1.718 µmol/g of CH_4 after a 6 h reaction, which is 1.6 and 14 times greater than the O_2-TiO_2 sample and the commercially available Degussa P25, respectively. This is because of the synergistic interaction between the excess oxygen defect in the O_2-TiO_2 sample, which is responsible for the sample's ability to absorb visible light, and the improved charge separation and surface reaction of excitons at the proximal junction of GO-OTiO_2 heterostructures (Tan et al. 2015).

TABLE 5.2

Table showing the fabrication methods and reaction conditions employed with different photocatalysts

Sl no	Catalyst	Fabrication method	Irradiation source	Major products	Yield/evolution rate	References
1	CdS-TiO$_2$-60 nanosheet	Photodeposition method	300W Xe arc lamp	CO CH$_4$	CO = 3.64 µmol/g/h CH$_4$ = 0.54 µmol/g/h	(Pan and Xu 2015)
2	Graphene/ g-C$_3$N$_4$	Facile One-pot impregnation-Thermal reduction method	15W daylight bulb	CH$_4$	CH$_4$ = 5.87 µmol/g	(Ong et al. 2015)
3	GO-OTiO$_2$	Wet chemical impregnation method	15W daylight bulb	CH$_4$	CH$_4$ = 1.718 µmol/g	(Tan et al. 2015)
4	BiOBr-4 nanosheet	Facile hydrothermal method	Xe lamp with intensity 0.21 W/cm^2	CO	CO = 4.45 µmol/g/h	(Wu et al. 2017)
5	Bi$_2$O$_2$(OH)$_{1+x}$(NO3)$_{1-x}$	Ultrasound-assisted top-down route	300W Xe lamp	CO	CO = 16.7 µmol/g	(Li et al. 2022)
6	Bi$_2$O$_2$(NO$_3$)(OH)/g-C$_3$N$_4$	Facile electrostatic self-assembly method	300W Xe lamp	CO	CO = 14.84 µmol/g/h	(Liu et al. 2022)
7	BP@g-C$_3$N$_4$	Electrostatic attraction approach	300W Xe lamp	CO	CO = 6.54 µmol/g/h	(Han et al. 2018)
8	rGO-MoS$_2$/PPy	Hydrothermal method	300W Xe lamp	CO CH$_4$ H$_2$	CO = 3.95 µmol/g/h CH$_4$ = 1.50 µmol/g/h H$_2$ = 4.19 µmol/g/h	(Kumar et al. 2020)
9	S-C-S MoS$_2$/SnS$_2$/r-GO	Solvothermal method	8W mercury lamp	CO CH$_4$	CO = 68.53 µmol/g/h CH$_4$ = 50.55 µmol/g/h	(Yin et al. 2019)
10	MoS$_2$/h-BN	Hydrothermal method	20W LED spotlight	CH$_3$OH	CH$_3$OH = 5994 µmol/g	(Kumari et al. 2020)
11	α-Fe$_2$O$_3$/MoS$_2$	In situ growth hydrothermal method	Visible light	CH$_4$ CH$_3$OH	CH$_4$ = 121 µmol/g/h CH$_3$OH = 41 µmol/g/h	(Zhao et al. 2020)
12	TiO$_2$ NSs/TCPP	Self-assembly approach	Visible light	CO	CO = 141.74 µmol/g/h	(Gao et al. 2019)
13	Ag$_2$S-In$_2$S$_3$	Ion-exchange strategy	300W Xe Lamp	CO CH$_4$	CO = 25.1 µmol/g/h CH$_4$ = 20 µmol/g/h	(Shao et al. 2021)
14	SnS$_2$/Au/g-C$_3$N$_4$	Facile thermal treatment and chemical deposition method	Solar simulator	CO CH$_4$	CO = 93.81 µmol/g/h CH$_4$ = 74.98 µmol/g/h	(Yin et al. 2021)

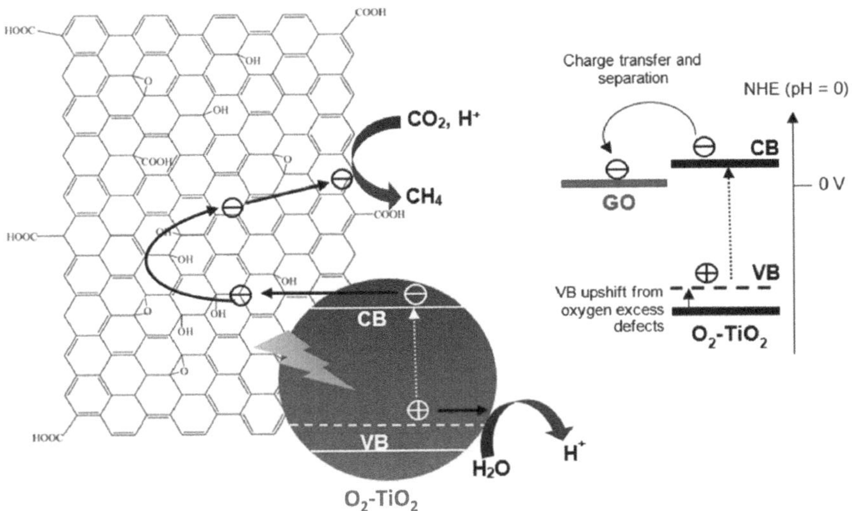

FIGURE 5.4 The proposed mechanism of charge transfer and separation process of GO-OTiO$_2$ hybrid nanocomposite for the photocatalytic conversion of CO$_2$ to CH$_4$. Reproduced with permission from Ref. (Tan et al. 2015).

One of the most significant 2D nanomaterials for the photoreduction of CO$_2$ to solar fuel is bismuth oxyhalogen (BiOX) material, where X = Cl, Br, or I. The BiOX materials can absorb UV and visible (Vis) light. For understanding the influence of the (001) facet of BiOBr in the photocatalytic conversion of CO$_2$ to CO, Dan Wu and co-workers produced BiOBr nanosheets using a simple hydrothermal technique. Nitric acid solutions of 0, 0.5, 1, and 4 M were selected for this preparation. The precursors for bismuth and bromine in this synthesis were bismuth(III) nitrate pentahydrate (Bi(NO$_3$)$_3$.5H$_2$O) and potassium bromide (KBr). According to the XPS results, BiOBr-0 had a conduction band maximum (CBM) of −0.29 eV, while BiOBr-4 had a CBM of −0.62 eV. The sample BiOBr-4 played an important role in converting CO$_2$ to CO, as shown by the CO$_2$/CO redox potential value of −0.53eV. The BiOBr-4 nanosheets were capable of producing electrons with effective charge carrier separation and charge transfer, which led to the conversion of CO$_2$ to CO with a yield of 4.45 μmol/g/h (Wu et al. 2017).

Tongyao Liu and co-workers synthesized and studied bismuth-based photocatalysts, which were incorporated into g-C$_3$N$_4$ for higher photocatalytic activity in CO$_2$ conversion to solar fuel generation (Figure 5.5). They successfully generated Bi$_2$O$_2$(NO$_3$)(OH)/g-C$_3$N$_4$ hybrid material (BON/CN) by a simple electrostatic self-assembly technique. The photocatalytic reduction activities of BON, g-C$_3$N$_4$, and BON/CN were separately studied and, among them, the BON/CN composite showed the highest activity toward CO formation, with a yield of 14.84 μmol/g/h under simulated solar light irradiation. This result is 15.6 times and 4.5 times greater than bare BON and g-C$_3$N$_4$ catalysts, respectively (Liu et al. 2022).

Chunqiu Han and co-workers researched the significance of metal-free semiconductor nanomaterial photocatalysts for the CO$_2$ photoreduction reaction. Through a

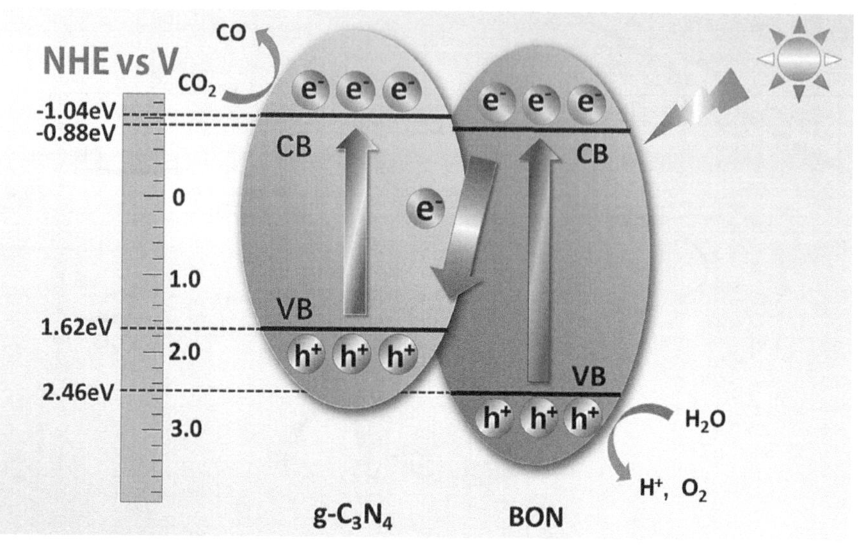

FIGURE 5.5   The proposed mechanism of Bi$_2$O$_2$(NO$_3$)(OH)/g-C$_3$N$_4$ Z-scheme heterostructure. Reproduced with permission from Ref. (Liu et al. 2022).

straightforward electrostatic attraction mechanism, they created a black phosphorous (BP) quantum dot/g-C$_3$N$_4$ nanocomposite. One of the key characteristics of g-C$_3$N$_4$ is its negative conduction band potential (−1.12eV), which is very successful in photoreduction processes. BP is a 2D-layered material similar to graphene with superior electrical and optical properties and also has a high response in the solar spectrum (Han et al. 2018).

MoS$_2$ is a promising material for making solar fuel through effective CO$_2$ photoreduction. Neeraj Kumar and co-workers synthesized polypyrrole-promoted rGO-MoS$_2$ hybrid material for the study of the photocatalytic conversion of CO$_2$ into CO, CH$_4$, and H$_2$. GO was successfully produced using the modified Hummer's method, and the hydrothermal approach was used to prepare the combination of rGO-MoS$_2$. A conducting polymer called polypyrrole was added to the rGO-MoS$_2$ sample to create rGO-MoS$_2$/Ppy nanocomposites. The FT-IR spectra (400–4000cm^{-1}) were used to observe the functional groups present in the nanocomposite sample (Kumar et al. 2020).

Shikang Yin and co-workers have created another hybrid nanomaterial based on MoS$_2$. They created a unique S-C-S MoS$_2$/SnS$_2$/r-GO heterojunction for the photocatalytic reduction of CO$_2$ by a solvothermal method. Reactive oxygen species in MoS$_2$, MoS$_2$/SnS$_2$, and MoS$_2$/SnS$_2$/r-GO were analyzed by electron spin resonance spectroscopy using ultraviolet-visible (UV-Vis) light illumination. GC was used to identify the gas-phase products produced during CO$_2$ photoreduction. The hybrid material contains varying weight percentages of r-GO and SnS$_2$, and the products were obtained under UV light (Yin et al. 2019). A MoS$_2$/h-BN nanoplatelet hybrid material has been developed for the study of CO$_2$ photoconversion under visible light irradiation. Sangita Kumari and co-workers incorporated the MoS$_2$ nanostructure into a hexagonal boron nitride (h-BN) nanoplatelet by a hydrothermal method.

The specific product obtained by the conversion of CO_2 is CH_3OH with a yield of 5994 µmol/g, which is 3.8 times higher than pristine MoS_2 under visible light irradiation (Kumari et al. 2020).

Yunxia Zhao and co-workers fabricated visible light active flower-like MoS_2 with Fe-S bonds by decoration of nano ferric oxide (Fe_2O_3) for efficient photoreduction of CO_2. Here, sodium molybdate dihydrate and $FeCl_3$ were used as the precursors for MoS_2 and α-Fe_2O_3. The separation and transportation of photoinduced carriers were facilitated by the stable heterostructure formed by Fe-S links connecting α-Fe_2O_3 and MoS_2. The major products when irradiated with visible light were CH_4 and CH_3OH (Zhao et al. 2020).

Having a UV sensitivity in the solar spectrum, TiO_2 is a frequently used semiconductor material. To explore more synthetic approaches, researchers are now examining how to engage the visible region. There has been a lot of interest recently in the production of nanohybrids using different organic materials and TiO_2. Hongyi Gao and co-workers successfully synthesized a highly efficient TiO_2 nanosheet/ tetra(4-carboxy phenyl)porphyrin hybrid (TiO_2 NSs/TCPP) for CO_2 photoreduction application. Comparison with bare TiO_2 NSs revealed that the hybrid material had the highest photocatalytic activity. In this study, TCPP acts as a light-harvesting unit to generate light-induced e^--h^+ pairs. The UV-Vis diffuse reflectance spectroscopy proved the successful impregnation of TCPP on TiO_2 nanosheet, and the FT-IR study investigated the interaction between TiO_2 nanosheet and TCPP. Different compositions of TiO_2NSs/TCPP were prepared using different amounts of TCPP, and the photocatalytic reduction mechanism was studied under visible light, and it was observed that after 12 h irradiation, the two-electron reduction product CO had greater accessibility. When the loading of TCPP reached 11.5%, the composite material showed greater catalytic performance for CO, and the yield obtained was 141.74 µmol/g/h, which is 37 times higher than the bare TiO_2 NSs material (Gao et al. 2019).

In the photocatalytic conversion of CO_2 to value-added compounds, hybrid photocatalysts including noble metal sulfide have attracted a lot of interest. Weiwei Shao and co-workers employed an ion exchange strategy to prepare silver sulfide- indium sulfide (Ag_2S-In_2S_3) 2D in-plane layered heterostructure photocatalyst (Shao et al. 2021). Density functional theory (DFT) calculations confirm electron transfer from Ag_2S to In_2S_3. Under light irradiation, the generated electron from the CBM of In_2S_3 migrated to the valence band maxima of the Ag_2S atomic layer, constituting the band structure that resembles the Z-scheme. The Ag_2S-In_2S_3 atomic layer enabled the intermediate CHO*, which is responsible for the desired product. According to this study, the hybrid material Ag_2S-In_2S_3 converts CO_2 to CH_4 with a yield of 20 µmol/ g/h, which is 16.7 times higher than the bare In_2S_3 atomic layer.

Noble metal-embedded heterostructure photocatalysts have great interest in this area. Shikang Yin and co-workers developed a SnS_2/Au/g-C_3N_4 ternary material for photoreduction application prepared by combining a simple thermal treatment and a chemical deposition technique (Figure 5.6). The Au nanoparticle improved the conductivity between the SnS_2/g-C_3N_4 structure and the Z-scheme mechanism, which facilitated the photoreduction process where Au provided the charge carrier transport (Yin et al. 2021).

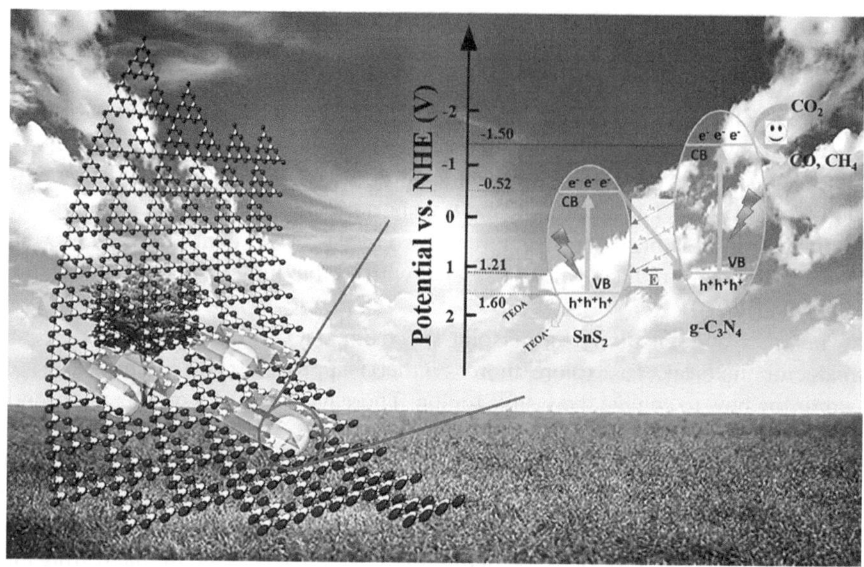

FIGURE 5.6 Schematic representation of energy band location and charge separation mechanism of $SnS_2/Au/g\text{-}C_3N_4$ heterostructure under solar light irradiation. Reproduced with permission from Ref. (Yin et al. 2021).

5.5 REACTOR DESIGN FOR CO_2 PHOTOREDUCTION

The development of suitable photoreactors for CO_2 photoreduction is currently of great interest. A favorable photoreactor design should allow for increased photon absorption and effective utilization of the e⁻-h⁺ pairs for superior photoreduction operation. Quartz, Pyrex, glass, and stainless steel are the materials used to make solar photoreactors. The photoreactors are generally categorized as slurry, fixed bed, and membrane photoreactors depending on their operational mode, type of bed, number of phases involved, membrane involved, and kind of light source (Figure 5.7) (Khan and Tahir 2019). This section discusses some of the solar photoreactors for the conversion of CO_2 into value-added chemicals.

To explore the photoreduction of CO_2 into CO and CH_4, Muhammad Tahir and co-workers employed indium-doped TiO_2 nanoparticles in a microchannel monolith photoreactor (Tahir and Amin 2013). The catalyst is uniformly coated as a thin layer on the surface of the monolith, which has a large number of cavities and permits strong surface interaction with radiation. The geometry of the monolith and operational parameters are the most important requirements for achieving maximum yield in solar reactors. The reactor in this investigation is a 150 cm³ stainless-steel cylindrical vessel with a length of 5.5 cm (Figure 5.8). The monoliths were obtained commercially and have dimensions of 6 cm in diameter, 0.5–10 cm in length, and 100 and 400 channels per square inch. A 200W mercury lamp is used as the UV light source that emits non-collimated light with a maximum intensity of 252 nm.

In another study, Muhammad Tahir used a continuous-flow monolith photoreactor to photo-reduce carbon dioxide into solar fuels using a nanocatalyst made of

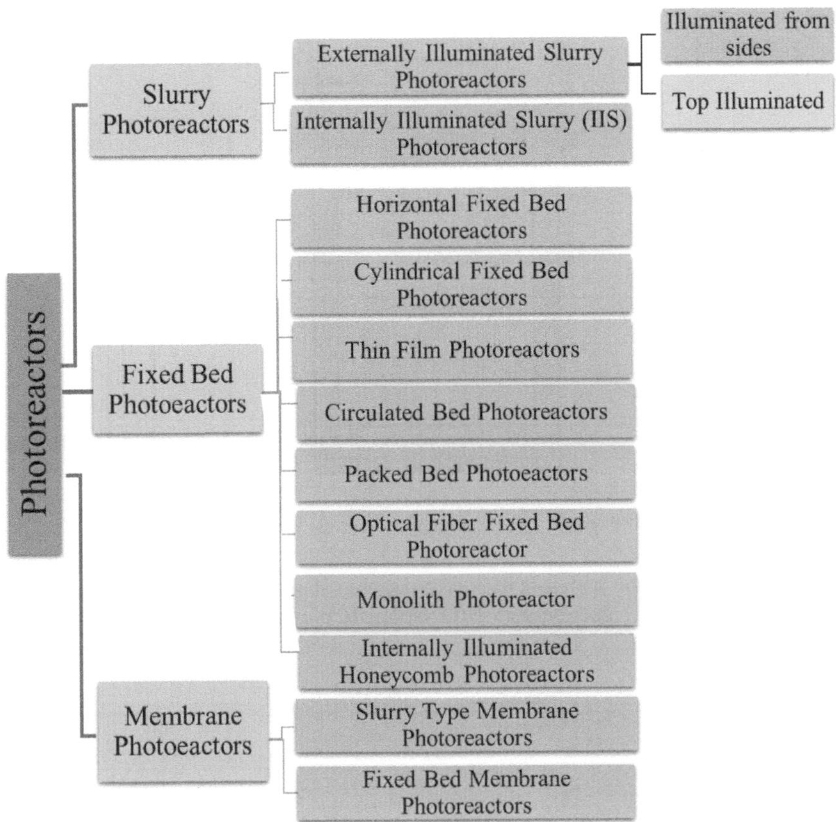

FIGURE 5.7 Classification of different types of photoreactors for CO_2 photoreduction. Reproduced with permission from Ref. (Khan and Tahir 2019).

Fe/TiO$_2$ distributed in montmorillonite (Tahir 2018). In this work, the stability of nanocomposite catalysts was compared using a cell type and monolith photoreactor, and the catalyst's life was also determined. It is made up of a cylindrical tank with a total volume of 150 cm^3 and a 200W Hg lamp as the light source, with a maximum intensity of 150 mW/cm^2 at a wavelength of 252 nm (Figure 5.9). A suitable amount of catalyst is uniformly suspended inside the cell-type photoreactors, while in monolith photoreactors the catalyst is uniformly coated as a layer on the surface of the monolith. A continuous flow of feed gas (CO_2 and H_2, purity = 99.99%) at a total flow rate of 20 mL/min was used to purge the reactor before the experiment began. All of the experiments were conducted at atmospheric pressure using the continuous flow method. To analyze the products, a GC outfitted FID and a thermal conductivity detector was utilized.

With the help of co-workers, Jeong Yeon Do constructed a batch-type photoreactor for the photoconversion of CO_2 into CH_4, utilizing metal-incorporated TiO$_2$/basalt fiber films as the catalyst (Do et al. 2016). Iron, cobalt, nickel, and copper were the different metals used in this study. The reactor consists of a 60.0 mL rectangular

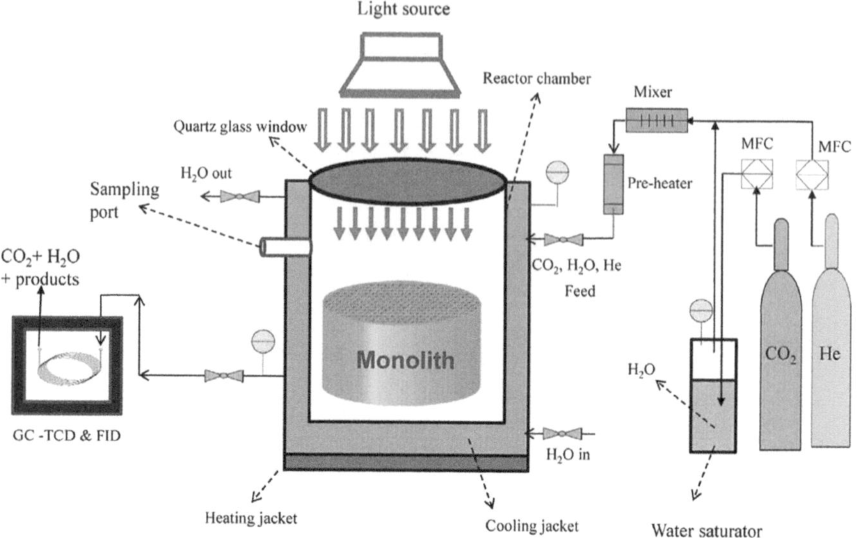

FIGURE 5.8 Schematic representation of experimental setup using monolith photoreactor for photocatalytic CO_2 reduction with H_2O vapors. Reproduced with permission from Ref. (Tahir and Amin 2013).

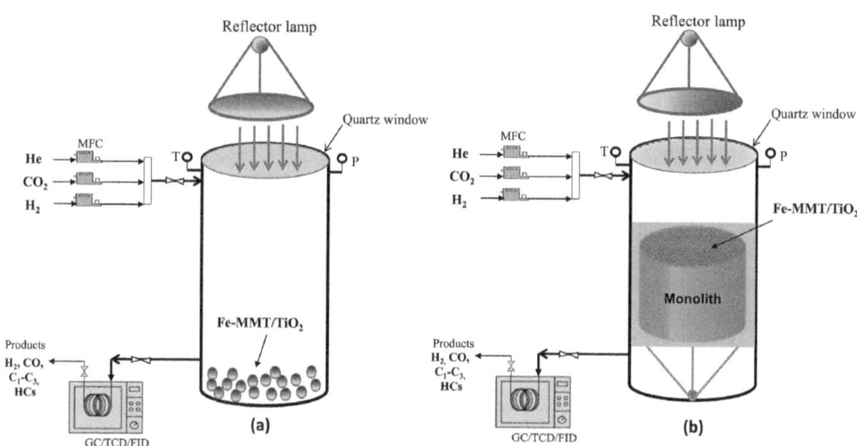

FIGURE 5.9 Schematic representation of the experimental setup: (a) cell photoreactor, (b) monolith photoreactor. Reproduced with permission from Ref. (Tahir 2018).

quartz cell as its entire capacity. The top of the reactor was covered with a window made of 1.0 mm thick quartz glass to effectively convey the 365 nm-wavelength irradiations from two 6.0-W/cm² mercury lamps. The reaction was kept at a temperature and pressure of 313 K and 1.0 atm, respectively. The reactor was purged with a CO_2 and helium mixture for an hour before the experiment started. A fixed CO_2:H_2O ratio of 1:2 was used (Figure 5.10).

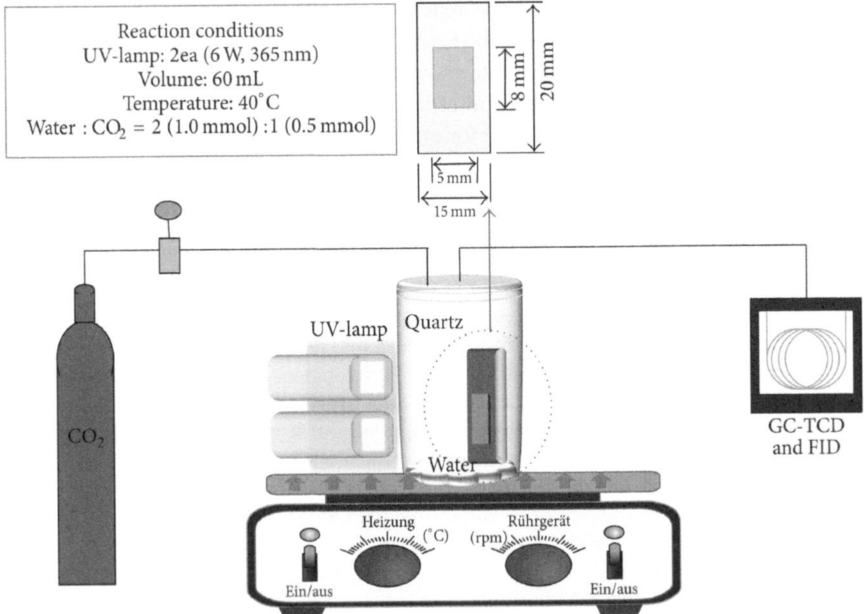

FIGURE 5.10 Schematic representation of batch-type photoreactor for photoreduction of CO_2 into CH_4. Reproduced with permission from (Do et al. 2016).

5.6 CONCLUSIONS, CHALLENGES, AND FUTURE PERSPECTIVES

Among the most efficient and environmentally friendly ways of combating climate change is the use of anthropogenic CO_2 to produce various value-added chemicals and fuels. Even though considerable progress has been achieved in effectively utilizing solar energy for the conversion of anthropogenic CO_2 to value-added chemicals or fuels with the help of various catalysts, the viability of these catalytic systems in practical uses or on a commercial scale can be improved by a large dimension. Recent advances in the area of 2D nanomaterials for photocatalytic CO_2 reduction are discussed in this chapter. As photocatalysts for CO_2 utilization, 2D nanomaterials seem to be of great interest due to their inherent properties, such as surface area, porosity, and crystal facet distribution, among others. At the same time, the synthesis of 2D nanomaterials on a commercial scale with uniform morphology and size is still a challenge that needs to be addressed.

The photogenerated carrier separation and the kinetics of the surface reaction are the two main bottlenecks for CO_2 utilization using photocatalysts. Studies show that heterojunctions can assist with charge separation, while surface defects can help with adsorption and activation. Thus, to overcome the key challenges of charge separation and reaction kinetics faced by the photocatalysts, the surface defects and heterojunction combinations can be optimized to get the maximum result, which is a promising method. Also, while designing the 2D nanomaterial-based photocatalyst, it is important to keep in mind to follow an environmentally benign, green, sustainable, and scalable protocol. Understanding the reaction pathways during the DFT modeling will help in desigming the photocatalyst with high efficiency and activity. Upcoming research in this area may concentrate more on the design and development of photocatalysts with improved activity and efficiency, being simple and environmentally friendly, cost-effective, and selective for efficient CO_2 utilization.

REFERENCES

Aizebeokhai, A. Phillips. 2009. Global warming and climate change: realities, uncertainties and measures. *International Journal of Physical Sciences* 4 (13):868–879.

Al Jitan, Samar, Giovanni Palmisano, and Corrado Garlisi. 2020. Synthesis and surface modification of TiO_2-based photocatalysts for the conversion of CO_2. *Catalysts* 10 (2):227.

Ali, Shahzad, Abdul Razzaq, and Su-Il In. 2019. Development of graphene based photocatalysts for CO_2 reduction to C1 chemicals: a brief overview. *Catalysis Today* 335:39–54.

Amani, Matin, Der-Hsien Lien, Daisuke Kiriya, et al. 2015. Near-unity photoluminescence quantum yield in MoS_2. *Science* 350 (6264):1065–1068.

Ameta, Rakshit, Meenakshi S. Solanki, Surbhi Benjamin, and Suresh C. Ameta. 2018. Photocatalysis. In *Advanced Oxidation Processes for Waste Water Treatment*, edited by Suresh C. Ameta, and Rakshit Ameta: Elsevier.

Angizi, Shayan, Sayed Ali Ahmad Alem, Mahdi Hasanzadeh Azar, et al. 2022. A comprehensive review on planar boron nitride nanomaterials: From 2D nanosheets towards 0D quantum dots. *Progress in Materials Science* 124:100884.

Bera, R., S. Kundu, and A. Patra. 2015. 2D hybrid nanostructure of reduced graphene oxide-CdS nanosheet for enhanced photocatalysis. *ACS Applied Materials & Interfaces* 7 (24):13251-9.

Chen, X., Y. Zhou, Q. Liu, Z. Li, J. Liu, and Z. Zou. 2012. Ultrathin, single-crystal WO_3 nanosheets by two-dimensional oriented attachment toward enhanced photocatalystic reduction of CO_2 into hydrocarbon fuels under visible light. *ACS Applied Materials & Interfaces* 4 (7):3372-7.

Chen, Xi, and Fangming Jin. 2019. Photocatalytic reduction of carbon dioxide by titanium oxide-based semiconductors to produce fuels. *Frontiers in Energy* 13 (2):207–220.

Cieśla, Paweł, Przemysław Kocot, Piotr Mytych, and Zofia Stasicka. 2004. Homogeneous photocatalysis by transition metal complexes in the environment. *Journal of Molecular Catalysis A: Chemical* 224 (1–2):17–33.

Dadashi-Silab, Sajjad, Sean Doran, and Yusuf Yagci. 2016. Photoinduced electron transfer reactions for macromolecular syntheses. *Chemical Reviews* 116 (17):10212–10275.

Di Sacco, Alice, Kate A. Hardwick, David Blakesley, et al. 2021. Ten golden rules for reforestation to optimize carbon sequestration, biodiversity recovery and livelihood benefits. *Global Change Biology* 27 (7):1328–1348.

Dincer, Ibrahim 2000. Renewable energy and sustainable development: a crucial review. *Renewable and Sustainable Energy Reviews* 4 (2):157–175.

Do, Jeong Yeon, Byeong Sub Kwak, Sun-Min Park, and Misook Kang. 2016. Effective carbon dioxide photoreduction over metals (Fe-, Co-, Ni-, and Cu-) incorporated TiO_2/basalt fiber films. *International Journal of Photoenergy* 2016:1–12.

Dong, Feng, Ying Wang, Bin Su, Yifei Hua, Yuanqing Zhang, and Recycling. 2019. The process of peak CO_2 emissions in developed economies: a perspective of industrialization and urbanization. *Resources, Conservation and Recycling* 141:61–75.

Fox, Marye Anne, and Maria T. Dulay. 1993. Heterogeneous photocatalysis. *Chemical Reviews* 93 (1):341–357.

Gao, Hongyi, Junyong Wang, Mengyi Jia, et al. 2019. Construction of TiO_2 nanosheets/tetra (4-carboxyphenyl) porphyrin hybrids for efficient visible-light photoreduction of CO_2. *Chemical Engineering Journal* 374:684–693.

Güney, Taner 2019. Renewable energy, non-renewable energy and sustainable development. *International Journal of Sustainable Development & World Ecology* 26 (5):389–397.

Han, Chunqiu, Jue Li, Zhaoyu Ma, et al. 2018. Black phosphorus quantum dot/g-C_3N_4 composites for enhanced CO_2 photoreduction to CO. *Science China Materials* 61 (9):1159–1166.

Hu, Boxun, Curtis Guild, and Steven L. Suib. 2013. Thermal, electrochemical, and photochemical conversion of CO_2 to fuels and value-added products. *Journal of CO_2 Utilisation* 1:18–27.

Hu, Zehua, Zhangting Wu, Cheng Han, Jun He, Zhenhua Ni, and Wei Chen. 2018. Two-dimensional transition metal dichalcogenides: interface and defect engineering. *Chemical Society Reviews* 47 (9):3100–3128.

Jelmy, E.J., Nishanth Thomas, Dhanu Treasa Mathew, et al. 2021. Impact of structure, doping and defect-engineering in 2D materials on CO_2 capture and conversion. *Reaction Chemistry & Engineering* 6 (10):1701–1738.

Jiang, Jie, Tao Xu, Junpeng Lu, Litao Sun, and Zhenhua Ni. 2019. Defect engineering in 2D materials: precise manipulation and improved functionalities. *Research* 2019:1–14.

Khan, Azmat Ali, and Muhammad Tahir. 2019. Recent advancements in engineering approach towards design of photo-reactors for selective photocatalytic CO_2 reduction to renewable fuels. *Journal of CO_2 Utilisation* 29:205–239.

Khan, Khalid, Chi-Wei Su, Ran Tao, and Lin-Na Hao. 2020. Urbanization and carbon emission: causality evidence from the new industrialized economies. *Environment, Development and Sustainability* 22 (8):7193–7213.

Kočí, Kamila, Han Dang Van, Miroslava Edelmannová, Martin Reli, and Jeffrey C. S. Wu. 2020. Photocatalytic reduction of CO_2 using Pt/C_3N_4 photocatalyts. *Applied Surface Science* 503:144426.

Kumar, B., M. Llorente, J. Froehlich, T. Dang, A. Sathrum, and C. P. Kubiak. 2012. Photochemical and photoelectrochemical reduction of CO_2. *Annual Review of Physical Chemistry* 63:541–569.

Kumar, Neeraj, Santosh Kumar, Rashi Gusain, Ncholu Manyala, Salvador Eslava, and Suprakas Sinha Ray. 2020. Polypyrrole-promoted rGO–MoS_2 nanocomposites for enhanced photocatalytic conversion of CO_2 and H_2O to CO, CH_4, and H_2 products. *ACS Applied Energy Materials* 3 (10):9897–9909.

Kumari, Sangita, Rashi Gusain, Anurag Kumar, Nilesh Manwar, Suman L. Jain, and Om P. Khatri. 2020. Direct growth of nanostructural MoS_2 over the h-BN nanoplatelets: an efficient heterostructure for visible light photoreduction of CO_2 to methanol. *Journal of CO_2 Utilisation* 42:101345.

Li, Yan-Yang, Yue-Yue Li, Xu Liang, et al. 2022. Atomically thin $Bi_2O_2(OH)_{1+x}(NO_3)_{1-x}$ nanosheets with regulated surface composition for enhanced photocatalytic CO_2 reduction. *ACS Applied Nano Materials* 5 (5):7019–7028.

Lin, Jingkai, Wenjie Tian, Huayang Zhang, Xiaoguang Duan, Hongqi Sun, and Shaobin Wang. 2020. Graphitic carbon nitride-based Z-scheme structure for photocatalytic CO_2 reduction. *Energy & Fuels* 35 (1):7–24.

Lingampalli, S. R., M. M. Ayyub, and C. N. R. Rao. 2017. Recent progress in the photocatalytic reduction of carbon dioxide. *ACS Omega* 2 (6):2740–2748.

Liu, Tongyao, Lin Hao, Liqi Bai, et al. 2022. Z-scheme junction $Bi_2O_2(NO_3)(OH)/g$-C_3N_4 for promoting CO_2 photoreduction. *Chemical Engineering Journal* 429:132268.

López-Carr, David, and Jason Burgdorfer. 2013. Deforestation drivers: population, migration, and tropical land use. *Environment, Development and Sustainability* 55 (1):3–11.

Mao, Jin, Kan Li, and Tianyou Peng. 2013. Recent advances in the photocatalytic CO_2 reduction over semiconductors. *Catalysis Science & Technology* 3 (10):1253–1260.

McCarthy, M. P., M. J. Best, and R. A. Betts. 2010. Climate change in cities due to global warming and urban effects. *Geophysical Research Letters* 37 (9):L0970.

Mills, Andrew, and Stephen Le Hunte. 1997. An overview of semiconductor photocatalysis. *Journal of Photochemistry and Photobiology A: Chemistry* 108 (1):1–35.

Mishra, Ashok K, Vijay P Singh, and Sharad K Jain. 2010. Impact of global warming and climate change on social development. *Journal of Social Welfare and Family Law* 26 (2–3):239–260.

Nguyen, T. P., D. L. T. Nguyen, V. H. Nguyen, et al. 2020. Recent advances in TiO_2-based photocatalysts for reduction of CO_2 to fuels. *Nanomaterials* 10 (2):337.

Ong, W. J., L. L. Tan, S. P. Chai, and S. T. Yong. 2015. Graphene oxide as a structure-directing agent for the two-dimensional interface engineering of sandwich-like graphene-g-C_3N_4 hybrid nanostructures with enhanced visible-light photoreduction of CO_2 to methane. *Chemical Communications* 51 (5):858–861.

Padmanabhan, Nisha T., Jayalatha Gopalakrishnan, and Honey John. 2022. Effect of reaction variables on facet-controlled synthesis of anatase TiO_2 photocatalysts. *Bulletin of Materials Science* 45 (2):1–15.

Padmanabhan, Nisha T., Nishanth Thomas, Jesna Louis, et al. 2021. Graphene coupled TiO_2 photocatalysts for environmental applications: a review. *Chemosphere* 271:129506.

Pan, Xiaoyang, and Yi-Jun Xu. 2015. Graphene-templated bottom-up fabrication of ultralarge binary CdS–TiO_2 nanosheets for photocatalytic selective reduction. *The Journal of Physical Chemistry C* 119 (13):7184–7194.

Rosso, Cristian, Giacomo Filippini, Alejandro Criado, Michele Melchionna, Paolo Fornasiero, and Maurizio Prato. 2021. Metal-free photocatalysis: two-dimensional nanomaterial connection toward advanced organic synthesis. *ACS Nano* 15 (3):3621–3630.

Roy, Soumyabrata, Arjun Cherevotan, and Sebastian C. Peter. 2018. Thermochemical CO_2 hydrogenation to single carbon products: scientific and technological challenges. *ACS Energy Letters* 3 (8):1938–1966.

Sato, S., S. Tanaka, K. I. Yamanaka, et al. 2021. Study of excited states and electron transfer of semiconductor-metal-complex hybrid photocatalysts for CO_2 reduction by using picosecond time-resolved spectroscopies. *Chemistry* 27 (3):1127–1137.

Shao, Weiwei, Shumin Wang, Juncheng Zhu, et al. 2021. In-plane heterostructured Ag_2S-In_2S_3 atomic layers enabling boosted CO_2 photoreduction into CH_4. *Nano Research* 14 (12):4520–4527.

Shit, S. C., I. Shown, R. Paul, K. H. Chen, J. Mondal, and L. C. Chen. 2020. Integrated nano-architectured photocatalysts for photochemical CO_2 reduction. *Nanoscale* 12 (46):23301–23332.

Sun, Lifei, Xingxu Yan, Jingying Zheng, et al. 2018. Layer-dependent chemically induced phase transition of two-dimensional MoS_2. *Nano Letters* 18 (6):3435–3440.

Tahir, Muhammad 2018. Photocatalytic carbon dioxide reduction to fuels in continuous flow monolith photoreactor using montmorillonite dispersed Fe/TiO_2 nanocatalyst. *Journal of Cleaner Production* 170:242–250.

Tahir, Muhammad, and NorAishah Saidina Amin. 2013. Photocatalytic CO_2 reduction and kinetic study over In/TiO_2 nanoparticles supported microchannel monolith photoreactor. *Applied Catalysis A-General.* 467:483–496.

Tan, Chaoliang, Xiehong Cao, Xue-Jun Wu, et al. 2017. Recent advances in ultrathin two-dimensional nanomaterials. *Chemical Reviews* 117 (9):6225–6331.

Tan, Lling-Lling, Wee-Jun Ong, Siang-Piao Chai, Boon Tong Goh, and Abdul Rahman Mohamed. 2015. Visible-light-active oxygen-rich TiO_2 decorated 2D graphene oxide with enhanced photocatalytic activity toward carbon dioxide reduction. *Applied Catalysis B: Environmental* 179:160–170.

Truc, Nguyen Thi Thanh, Long Giang Bach, Nguyen Thi Hanh, et al. 2019. The superior photocatalytic activity of Nb doped TiO_2/g-C_3N_4 direct Z-scheme system for efficient conversion of CO_2 into valuable fuels. *Journal of Colloid and Interface Science* 540:1–8.

Tu, Wenguang, Yong Zhou, and Zhigang Zou. 2013. Versatile graphene-promoting photocatalytic performance of semiconductors: basic principles, synthesis, solar energy conversion, and environmental applications. *Advanced Functional Materials* 23 (40):4996–5008.

Whipple, Devin T., and Paul J. A. Kenis. 2010. Prospects of CO_2 utilisation via direct heterogeneous electrochemical reduction. *The Journal of Physical Chemistry Letters* 1 (24):3451–3458.

Wu, Dan, Liqun Ye, Ho Yin Yip, and Po Keung Wong. 2017. Organic-free synthesis of {001} facet dominated BiOBr nanosheets for selective photoreduction of CO_2 to CO. *Catalysis Science & Technology* 7 (1):265–271.

Xiong, Jun, Pin Song, Jun Di, and Huaming Li. 2019. Ultrathin structured photocatalysts: a versatile platform for CO_2 reduction. *Applied Catalysis B: Environmental* 256:117788.

Yaashikaa, PR, P Senthil Kumar, Sunita J Varjani, and A Saravanan. 2019. A review on photochemical, biochemical and electrochemical transformation of CO_2 into value-added products. *Journal of CO_2 Utilisation* 33:131–147.

Yang, Lingming, Kausik Majumdar, Han Liu, et al. 2014. Chloride molecular doping technique on 2D materials: WS_2 and MoS_2. *Nano Letters* 14 (11):6275–6280.

Yang, Yanming, Ming Liu, Shitong Han, et al. 2021. Double-sided modification of TiO_2 spherical shell by graphene sheets with enhanced photocatalytic activity for CO_2 reduction. *Applied Surface Science* 537:147991.

Yin, S., J. Li, L. Sun, et al. 2019. Construction of heterogenous S-C-S MoS_2/SnS_2/r-GO heterojunction for efficient CO_2 photoreduction. *Inorganic Chemistry* 58 (22):15590–15601.

Yin, Shikang, Linlin Sun, Yaju Zhou, et al. 2021. Enhanced electron–hole separation in SnS_2/Au/g-C_3N_4 embedded structure for efficient CO_2 photoreduction. *Chemical Engineering Journal* 406:126776.

Yuan, Lan, and Yi-Jun Xu. 2015. Photocatalytic conversion of CO_2 into value-added and renewable fuels. *Applied Surface Science* 342:154–167.

Yui, T., A. Kan, C. Saitoh, K. Koike, T. Ibusuki, and O. Ishitani. 2011. Photochemical reduction of CO_2 using TiO_2: effects of organic adsorbates on TiO_2 and deposition of Pd onto TiO_2. *ACS Applied Materials & Interfaces* 3 (7):2594–2600.

Zaaba, NI, KL Foo, U Hashim, SJ Tan, Wei-Wen Liu, and CH Voon. 2017. Synthesis of graphene oxide using modified hummers method: solvent influence. *Procedia Engineering* 184:469–477.

Zhao, Yunxia, Wei Cai, Yunpeng Shi, et al. 2020. Construction of nano-Fe_2O_3-decorated flower-like MoS_2 with Fe–S bonds for efficient photoreduction of CO_2 under visible-light irradiation. *ACS Sustainable Chemistry & Engineering* 8 (33):12603–12611.

6 Engineering Semiconductor Nanomaterials for Air Purification

A. Aishu, S. Anjitha, K. Manoj and P. Periyat

6.1 INTRODUCTION

Almost 99% of the global population faces severe air pollution or is affected by polluted air. Indoor and outdoor air pollution daunts the human population, can cause respiratory and other diseases, and is an important source of morbidity and mortality. Continuous economic growth and industrialization are the major reasons for the increased rate of air pollution (Cao, Huang, and Zhang 2021). Pollutants of major concern include particulate matter, carbon monoxide (CO), Ozone (O_3), nitrogen dioxide (NO_2), sulfur dioxide (SO_2), and volatile organic compounds (VOCs) (Weon, Kim, and Choi 2018).

According to the reports from the World Health Organization (WHO) (Apte and Salvi 2016), around 2.4 billion people worldwide cook using open fires or inefficient stoves fueled by kerosene, biomass, and coal which generate harmful household air pollution. In 2020, 3.2 million deaths happened due to household air pollution and more than 2,00,000 were children under the age of 5. Indoor air pollution leads to some severe non-communicable diseases such as stroke, chronic obstructive pulmonary disease (COPD), ischemic heart disease, and lung cancer (Apte and Salvi 2016). Indoor air pollution is mainly caused by cooking and VOCs. The indoor VOCs are emitted from materials such as furniture, decorations, paints, building products, *etc.* The inhalation of these VOCs can cause various health problems such as allergies, nausea, headache, *etc.* (Chaudhary and Hellweg 2014; Chin et al. 2014). Long-term exposure to these can have carcinogenic, mutagenic, or teratogenic effects (Zhang, Tang, et al. 2011).

Outdoor air pollutants are also the greatest concern. It is one of the greatest risks to the environment and health. From the statistics from the WHO in 2019, almost 99% of the world's population was living in places where the WHO air-quality guidelines levels were not met. In the 2016 statistics, 4.2 million premature deaths occurred due to ambient air pollution in both cities and rural areas. The most terrific situation is that 91% of these premature deaths occurred in low- and middle-income countries, especially in Southeast Asia and the Western Pacific regions. Ambient air

DOI: 10.1201/9781003343899-6

pollution leads to lung cancer, COPD, other acute lower respiratory infections, and heart diseases.

Owing to their high capital costs, harsh conditions, and large area design, conventional air purifying methods are not suitable for the low concentrations of pollutants commonly encountered (Cao, Huang, and Zhang 2021). Conventional technologies used to control or purify air pollutants include electrostatic precipitators, wet scrubbers, fabric filters, and cyclonic precipitators (Pui, Chen, and Zuo 2014). In addition to this, VOCs are removed or recovered mainly by absorption, adsorption, and condensation. Besides, membrane separation and biofiltration also have a contribution. Furthermore, the adsorbed or recovered VOCs can be destroyed by combustion, catalytic, or biological oxidation processes (Cao, Huang, and Zhang 2021).

6.2 AMBIENT AIR PURIFICATION BY NANOPARTICLES

Recent developments in the world of nanotechnology give many opportunities for the remediation of air pollution. When compared with bulk materials, nanomaterials have larger surface areas and show different electronic, magnetic, and catalytic properties for pollutants and that is why they can be used for filtration and adsorption leading to air purification. Due to the special properties such as high surface area and high adsorption capacity shown by nanomaterials, they can be used for the purification of air pollutants which are even in low concentrations and have the advantage that nanomaterials for air purification only need moderate working conditions and they do not need any complicated instrumental setup (Cao, Huang, and Zhang 2021).

6.3 NANO TITANIA (TIO$_2$) FOR AIR PURIFICATION

Nano titania (TiO$_2$) is an important nanomaterial that has been widely investigated due to its high photocatalytic activity (Zhu et al. 2017). Recent studies on TiO$_2$ and its modifications (Weon, Kim, and Choi 2018; Zhang, Tang, et al. 2011; Zhu et al. 2017; Rao et al. 2019) show its potential for air purification. Nano titania (TiO$_2$) itself shows high photocatalytic activity, hence it has many applications in various fields and has been widely investigated because of its abundance, low material cost, chemical and photochemical stability, and strong photooxidative power (Cao, Huang, and Zhang 2021). TiO$_2$ has an integral role in air purification also. However, TiO$_2$ alone will have a lower photocatalytic degradation activity. TiO$_2$ deactivates during the continuous process of photocatalysis and this would diminish the commercial value as a practical air-purification technology. Hence, to improve its photocatalytic degradation efficiency, there is a need for modifying the nano TiO$_2$ using different methods of modification *via* heterojunction with other semiconductors and metals (Er, Pt, Au, and Ag) – and new photocatalyst materials can be synthesized (Weon, Kim, and Choi 2018).

Weon, Kim, and Choi (2018), Zhang, Tang et al. (2011), Zhu et al. (2017) and Rao et al. (2019) reported that the simple surface modification of TiO$_2$ prevents the catalyst deactivation. The degree of photocatalyst deactivation depends on various experimental conditions – mass and surface property, concentration, and kind of substrates, humidity level, O$_2$ concentration, and light intensity (Zahra, Seyed Morteza, and Aniseh Kafi 2019; Weon and Choi 2016). The mineralization of VOCs with accumulating recalcitrant intermediate leads to the deactivation of photocatalysts (Le Bechec et al. 2015).

6.3.1 Modified Nano TiO$_2$ for Air Purification

6.3.1.1 Dual Components Modified TiO$_2$ (Platinum and Fluoride)

Platinum (Pt)-loaded TiO$_2$ obtained by photo deposition method was surface fluorinated by soaking in the NaF solution for 30 min and dried under air to obtain the dual component modified TiO$_2$ (Weon, Kim, and Choi 2018). Surface fluorination that replaces the hydroxyl groups on the TiO$_2$ surface that facilitates the generation of mobile OH radicals thereby enhancing the photoinduced hydrophilicity and modulates water molecules adsorption. Further, it obstructs the adsorption of VOCs and their degradation intermediates. The OH radicals generated because of the selective transfer of holes to molecularly adsorbed H$_2$O rather than the transfer to the surface hydroxyl group. The selective hole transport to-adsorbed H$_2$O is caused by fluoride-induced depletion of surface hydroxyl groups. The more holes that can be employed for oxidation reactions since the holes that are transferred into mobile OH radicals are less available to the recombination with electrons present in the conduction band (Weon, Kim, and Choi 2018).

$$\text{Ti-OH} + \text{F}^- \leftrightarrow \text{Ti-F} + \text{OH}^-$$
$$\text{Ti-F (F-TiO}_2 \text{ surface)} + \text{H}_2\text{O}_{ad} + \text{h}_{vb}^+ \rightarrow \text{Ti-F} + \text{H}_+ + {}^{\bullet}\text{OH}_{mobile}$$
$$\text{Ti-OH (bare TiO}_2 \text{ surface)} + \text{h}_{vb}^+ \rightarrow \text{Ti-OH}^{\bullet}{}_{ad}$$
$$\text{Ti-OH}^{\bullet}{}_{ad} + \text{e}_{cb}^- \rightarrow \text{Ti-OH}$$

The surface platinum increases the charge carrier life span by making more holes that will react with adsorbed water molecules to generate mobile OH radicals. The photocatalytic VOC degradation efficiency is increased for Pt-deposited TiO$_2$ (Pt/TiO$_2$ and F-TiO$_2$/Pt) than bare TiO$_2$ (Figure 6.1). Platinum deposited on the TiO$_2$ surface separates charge pairs through the Schottky barrier formed at the interface of the TiO$_2$ surface and Platinum. The reactive radical species are formed due to the preferential precedence of rapid charge separation over slower interfacial charge transfer processes. Pt facilitated the efficient generation of reactive radical species by ensuring more charge carriers which resulted in effective charge seperation. When both platinum and fluorine species are present together (Figure 6.1), a larger number of mobile OH radicals are generated on the surface of F-TiO$_2$/Pt, and it enhances the photocatalytic activity of VOC degradation and the resistance against the catalyst deactivation during VOC degradation (Weon, Kim, and Choi 2018).

When combining both platinization and surface fluorination on TiO$_2$, a perfectly modified nano TiO$_2$ for the degradation of pollutant toluene is formed (Weon, Kim, and Choi 2018). The dual components modified TiO$_2$ (F-TiO$_2$/Pt) exhibits improved activity of remote photocatalytic oxidation which is mediated by mobile OH radicals and high resistance against the catalyst surface deactivation. When comparing with other photocatalysts such as TiO$_2$ alone, platinized TiO$_2$ (Pt/TiO$_2$), and

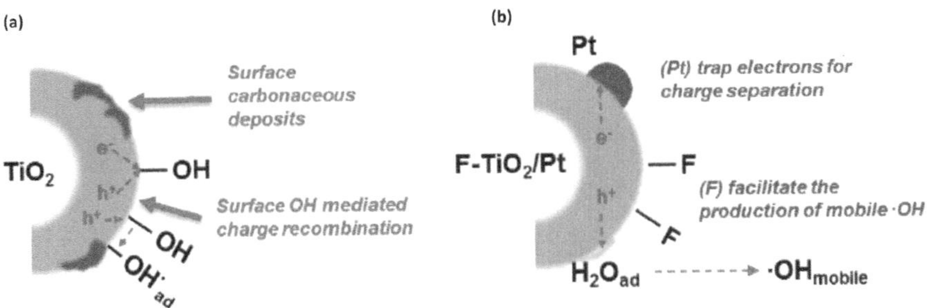

FIGURE 6.1   Schematic illustration of the photocatalytic reaction processes occurring on the surface of (a) bare TiO$_2$ and (b) F-TiO$_2$/Pt. Reproduced with permission from (Weon, Kim, and Choi 2018).

surface-fluorinated TiO$_2$ (F-TiO$_2$), this dual-component-modified TiO$_2$ attained a higher degradation efficiency and mineralization efficiency of VOCs (e.g. toluene) as well as have higher durability (Weon, Kim, and Choi 2018).

6.3.1.2   Nanocomposite of Ag–AgBr–TiO$_2$

Ag–AgBr–TiO$_2$ synthesized using a modified deposition–precipitation method produces a photoactive and durable catalyst for the gas-phase degradation of VOCs such as benzene and acetone (Figure 6.2). This modified catalyst of TiO$_2$ shows stable photocatalysis under both visible and UV light. When comparing the reaction of Ag–AgBr–TiO$_2$ in the presence of UV light and visible light, the improved reaction takes place in the presence of UV light because the intensity of hydroxyl radicals and superoxide radicals under the UV light is much higher than that of visible light. The photogenerated radicals enhance the gas-phase degradation of benzene and acetone. The loading of Ag–AgBr into TiO$_2$ helps to avoid the deactivation of TiO$_2$ during the degradation of VOCs (Zhang, Tang, et al. 2011).

Zhang, Tang, et al. (2011) reported that the primary photoactive components responsible for the degradation of gaseous phase benzene and acetone are Ag–AgBr based on the CO$_2$ production during the VOC degradation activity. TiO$_2$ acts only as an inert support. Because of the stable and inert chemical structure, benzene shows less degradation than acetone during photocatalytic degradation. The formation rate of hydroxyl radicals in Ag–AgBr–TiO$_2$ is very much higher than that of bare TiO$_2$, which aids in the higher visible light photocatalytic activity toward the degradation of VOCs like benzene and acetone (Zhang, Tang, et al. 2011).

6.3.1.3   Plasmonic Au/TiO$_2$

TiO$_2$-supported Au catalysts (Au/TiO$_2$) have great potential for VOC degradation due to the wide range and strong absorption of visible light *via* surface plasmon resonance of gold (Au) nanoparticles (Figure 6.3). Compared to photocatalysis, plasmonic photocatalysis has significant applications in air purification. In plasmonic photocatalysis, the synergy of noble metals such as gold, silver, *etc.* and semiconductor photocatalysts like TiO$_2$ are utilized and it brings notable changes to many aspects of VOC degradation by photocatalysis (Zhu et al. 2017).

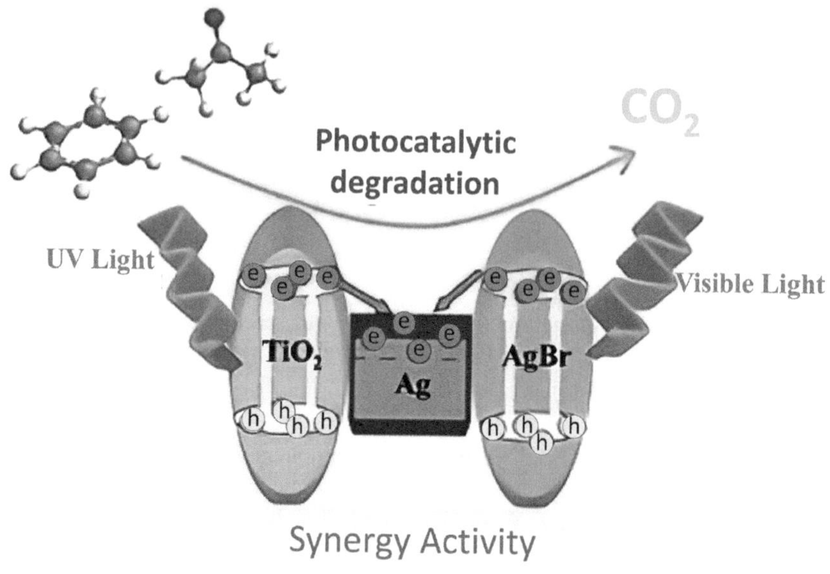

FIGURE 6.2   Diagram representing the degradation of benzene and acetone by Ag–AgBr–TiO$_2$ nanocomposite (Zhang, Tang, et al. 2011). Reproduced with permission from (Zhang, Tang, et al. 2011).

TiO$_2$ is modified by impregnating gold (Au) by using the modified impregnation method to fabricate a 4.5 nm-sized plasmonic Au/TiO$_2$ photocatalyst. Under wet air and visible light, plasmonic Au/TiO$_2$ exhibits very high photocatalytic activity for VOC degradation, that is, a complete conversion of the VOC–formaldehyde (almost 84%). Relative humidity plays a significant role in the photocatalytic oxidation of formaldehyde by the plasmonic Au/TiO$_2$. In the presence of high relative humidity, visible light contributes to enhanced formate oxidation and carbonate decomposition. Visible light enhancement promotes the decomposition of carbonate species that is formed during CO$_2$ adsorption (Zhu et al. 2017).

Conventional catalysis coupled with photocatalysis enabled the visible-light-mediated formaldehyde oxidation over plasmonic Au/TiO$_2$. Conventional catalysis involves the adsorption of formaldehyde on TiO$_2$ surface and nucleophilic attack by active oxygen species (oxygen molecules activated by water) at the interface between Au and TiO$_2$. The nucleophilic attack of electron-rich active oxygen species results in the formation of dioxymethylene (DOM), subsequent formate, carbonate, or CO$_2$ (Zhu et al. 2017). Zhu et al. (2017) reported that, in plasmonic photocatalysis, the hot electrons of gold come directly in contact with the conduction band of TiO$_2$ and are consumed by the reduction of an oxygen molecule to form superoxide radical species at the interface of Au and TiO$_2$. As a result, the cations migrate toward the Au–TiO$_2$ interface to promote the formaldehyde oxidation, forms hydroxyl radicals (oxidant) by water oxidation and initiate the photocatalytic reaction of VOCs degradation over plasmonic Au/TiO$_2$.

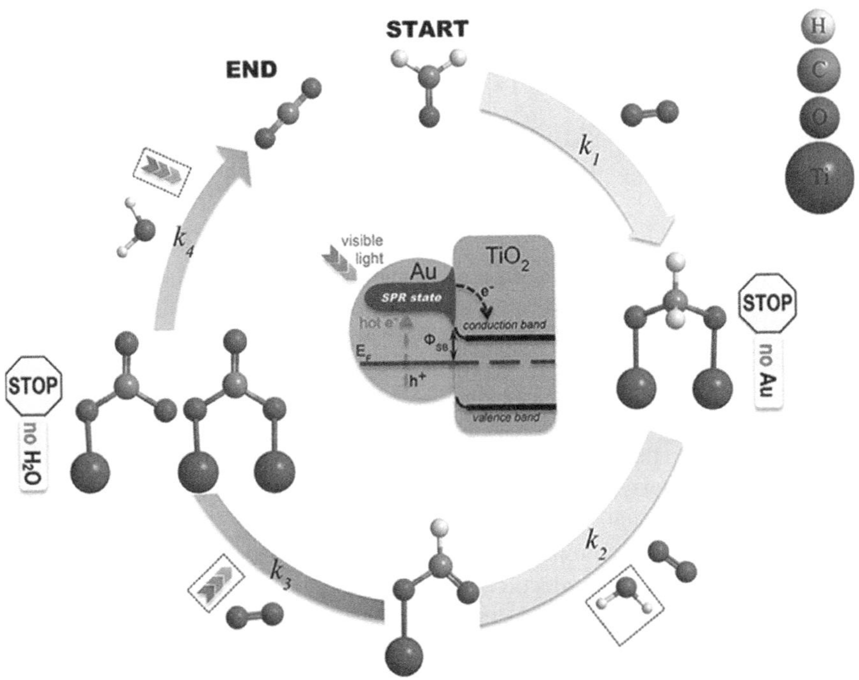

FIGURE 6.3 Diagram of a proposed pathway for removal of formaldehyde over Au/TiO_2 under visible light. Symbols of white, gray, red, and dark gray atom balls represent atoms of H, C, O, and Ti. The reaction proceeds from the start of formaldehyde to the end of CO_2, in which the kinetic parameters k_1, k_2, k_3, k_4 represent four conversion steps involving formaldehyde, intermediates (DOM, formate, and carbonate) and CO_2 (Zhu et al. 2017). Reproduced with permission from (Zhu et al. 2017).

6.3.1.4 Er^{3+} Doped TiO_2

Rao et al. reported the Er^{3+} doped TiO_2 synthesized using the sol-gel method, having a size of 9.5 nm against degradation of various VOCs such as acetaldehyde, o-xylene, and ethylene (Rao et al. 2019). The Er^{3+}-doping significantly increases the photocatalytic VOC degradation performance of TiO_2 in the removal of the above gaseous air pollutants. Also, the Er_2O_3 present helps in an increase of surface area which hence promotes the adsorption capacity of VOCs (Rao et al. 2019). Compared to the degradation principle of acetaldehyde and o-xylene, the degradation principles of ethylene are different. This can be attributed to the poor adsorption of ethylene molecules on the surface of the catalyst than acetaldehyde and o-xylene. As the three VOCs (acetaldehyde, o-xylene, and ethylene) have different molecular weight and structure, their degradation conditions are also different – at an optimum concentration of Er^{3+} ion (1.5%) exhibiting the best photocatalytic activity, the photodegradation efficiency and rate constant of acetaldehyde is higher than that of o-xylene and is minimum for ethylene. According to Rao et al. (2019), the adsorption capacity of the catalyst increased with an increase in Er^{3+}content for the same compound because of

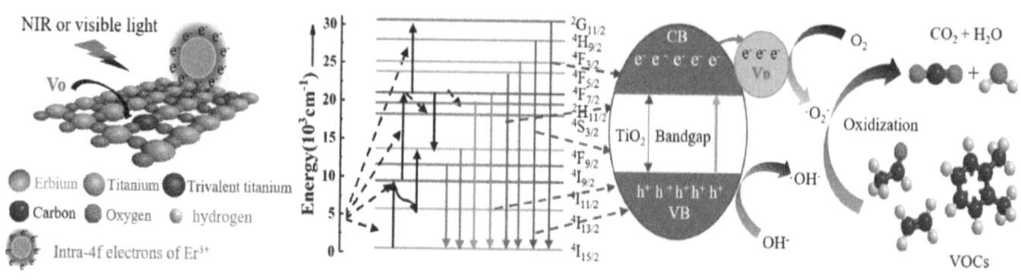

FIGURE 6.4 Diagram of the mechanism of intra-4f electron structure of Er^{3+} and Vo (Oxygen vacancy) promoting electron-to-electron transfer enhancement and up-conversion enhanced light energy utilization (Rao et al. 2019). Reproduced with permission from (Rao et al. 2019).

the increase in specific surface area. In the case of *o*-xylene degradation, it poisons TiO_2 which leads to the complete deactivation of TiO_2; however, the Er^{3+}-incorporated TiO_2 lattice has a delayed deactivation and it improved the degradation of o-xylene in a considerable amount. The photocatalytic performance of Er^{3+}-doped TiO_2 is not only related to adsorption properties but also related to the separation efficiency of the electron-hole pairs. The introduction of Er^{3+} in TiO_2 generates defects capturing photoinduced electrons and intra-4f electrons of Er^{3+}ion that accelerate the photogenerated charge-transfer transition between TiO_2 and Er^{3+} ion (Figure 6.4). This will increase the degradation activity generation and thus enhance the photocatalytic activity. This study showed that the oxygen radical formed during the reaction has a predominant role in the degradation of acetaldehyde and ethylene, however, in the case of o-xylene, hydroxyl radical has the dominant role in degradation (Rao et al. 2019).

6.4   2D SEMICONDUCTOR NANOMATERIALS FOR AIR PURIFICATION

Based on dimensions, nanomaterials are classified into zero-dimensional (0D), one-dimensional (1D), two-dimensional (2D), and three-dimensional (3D) nanoparticles. The same chemical compounds can exhibit different properties when they are configured into different dimensions (Zahra, Seyed Morteza, and Aniseh Kafi 2019). In recent years, 2D nanomaterials such as graphene and g-C_3N_4 stand out as excellent candidates in electronics, catalysts, energy storage facilities, sensors, solar cells, air purification, lithium batteries, nanocomposites, water purification, *etc.* due to their satisfactory properties (Zahra, Seyed Morteza, and Aniseh Kafi 2019). 2D nanomaterials are composed of thin layers and have a high aspect ratio, that is surface area to volume ratio. Due to the high aspect ratio, the presence of atoms on the surface is also high and these atoms have a different function than the internal atoms. The behavior of 2D nanomaterials depends on the number of surface atoms present.

As of now, few studies are available on 2D nanomaterials for air purification. Graphene is the most important and widely used 2D nanomaterial. Graphene possesses excellent properties such as good electronic properties, electron transfer capabilities, unprecedented impermeability, and high mechanical strength. Besides, properties such as excellent thermal and electrical conduction (Zahra, Seyed Morteza, and Aniseh Kafi 2019), high current density, chemical inertness, high thermal conductivity, optical transmittance, superficial hydrophobicity on a nanometer scale, *etc.* have made a great contribution toward the widespread use of graphene in diverse scientific areas (Zahra, Seyed Morteza, and Aniseh Kafi 2019). Along with graphene, its derivatives such as graphene oxide, reduced graphene oxide, and functionalized graphene have been the most studied 2D materials during the last decade for air purification. Recently, other ultrathin 2D nanomaterials are also of interest.

Graphitic carbon nitride (g-C_3N_4) is a 2D polymeric and layered semiconductor photocatalyst with a band gap of 2.7 e.V. Owing to its major advantages such as high stability, appealing electronic structure, and medium band gap, the g-C_3N_4 is multifunctional and has broad applications in the field of energy conservation and storage, contaminant degradation, carbon dioxide storage and reduction, catalysis, solar cells, and sensing (Wang, Wang, and Antonietti 2012; Zheng et al. 2012; Dong et al. 2013; Xiang, Yu, and Jaroniec 2011).

Synthesis of g-C_3N_4 is done by pyrolysis of nitrogen-rich precursors *via* polycondensation (Dong et al. 2014). Depending upon the condensation conditions and the types of precursors used, the texture, electronic structure, and performance of g-C_3N_4 can be changed (Wang, Wang, and Antonietti 2012; Zheng et al. 2012). Major precursors used for the synthesis of g-C_3N_4 include cyanamide, dicyanamide, trithiocyanuric acid, melamine, triazine, heptacene derivatives, *etc.* (Dong et al. 2013; Xiang, Yu, and Jaroniec 2011; Zhang, Sun, et al. 2011). Dong et al. (2012) reported urea and thiourea can also be used as a precursor for the same. To increase the reactivity in VOC degradation through photocatalysis, CO_2 reduction and selective synthesis templating, doping, heterostructure design, and post-functionalization are further techniques that can be used to modify the band structure and texture of g-C_3N_4 (Dong et al. 2014).

g-C_3N_4 has been synthesized to be a superior photocatalyst for air and aqueous pollutant degradation for air and water purification; however, its micro/nanostructures are needed to be enhanced (Dong et al. 2011, 2013; Xiang, Yu, and Jaroniec 2011; Zhang, Sun, et al. 2011; Yan, Li, and Zou 2009; Zhang et al. 2010; Bojdys et al. 2008). By implementing an enhancement in crystallinity and surface area, the photocatalytic activity of g-C_3N_4 can be improved. Enhancement of crystallinity will encourage the reduction of defects and inhibit charge carriers' recombination and at the same time, increased surface area could provide more active sites for adsorption and degradation of VOCs (Dong et al. 2014).

Dong et al. (2014) developed a method to engineer the micro/nanostructures of g-C_3N_4 from pyrolysis of thiourea and applied that for visible light photocatalytic air purification. This study showed that pyrolysis conditions have profound importance in changing the crystallinity, band structure, and surface area. With elevated pyrolysis temperature and prolonged pyrolysis time, the crystallinity and surface areas of g-C_3N_4 increase and thereby degradation of VOCs.

Dong et al. (2014) reported that the pyrolysis temperature has a direct effect on photocatalytic activity and NO removal. This study showed that g-C_3N_4 has a bandgap suitable for direct excitation by visible light and hence acts as an excellent photocatalyst for NO removal and thereby air purification. In the presence of g-C_3N_4 photocatalyst, the NO reacts with the photogenerated reactive radicals and produces the final product of HNO_3. During photocatalytic oxidation of NO, NO_2 is formed as a reaction intermediate. It was found that an enhancement in the surface areas and pore volumes with increased pyrolysis temperature and prolonged pyrolysis time because of a decrease in the fraction of NO_2 generated over g-C_3N_4 samples during irradiation.

The high surface areas and large pore volume promote the faster diffusion rate of the reaction intermediate over g-C_3N_4 which in turn promotes the oxidation of intermediate NO_2 to final NO^{3-}. Using water, the final oxidation products such as nitric acid or nitrate ions can be simply washed away so that air can be purified. During photocatalytic reaction, there occurs a conversion of NO to NO^{3-}; hence, the NO concentration decreases gradually. When the photocatalytic reaction reaches equilibrium, the NO concentration also reaches minimum (Dong et al. 2014).

$$NO + 2 \cdot OH \rightarrow NO_2 + H_2O$$
$$NO_2 + OH \rightarrow NO_3^- + H^+$$
$$NO + NO_2 + H_2O \rightarrow 2HNO_2$$
$$NO + \cdot O_2^- \rightarrow NO_3^-$$

When g-C_3N_4 is treated at high temperature for a long time, it enhances the crystallization; however, the enhancement of crystallization decreases the number of defects in the samples and thereby it is advantageous to reduce the recombination rate of photogenerated electrons and holes and improves the degradation of VOCs. Also, the nanosheet structure of g-C_3N_4 increases the mobility of photogenerated electrons along the nanosheets thus reducing the rate of electron-hole recombination. In addition to this the porous structure of the sample provides a large surface area for VOC adsorption and a large pore volume gives more active site for quick reactant diffusion and thereby higher degradation of VOC pollutants. The increased bandgap increases the redox ability of charge carriers formed through irradiation. All these factors favor the improved photocatalytic removal of VOCs using g-C_3N_4 samples obtained at elevated temperatures and prolonged time (Dong et al. 2014).

6.5 CONCLUSIONS

Both indoor and outdoor air pollution are the most important and sensational problems all over the world; hence it is important to find out the most suitable way to solve the concern. Research is going on worldwide to find out a sustainable solution to the daunting challenges of air pollution. Purification of air pollutants by nanotechnology is extensively being investigated; among the various nanomaterials reported, the semiconductor nanomaterials (*e.g.*, TiO_2) and 2D nanomaterials (*e.g.* graphene, g-C_3N_4) showed very good performance for the purification of air pollutants and give a viable solution for ambient air purification. TiO_2 and its derivative gave a viable solution to both indoor and outdoor air pollution and came in the form of viable technology. However, research in 2D nanomaterials still needs to be progressed. Most research in 2D nanomaterials is related to graphene and its derivatives such as graphene oxide, reduced graphene oxide and functionalized graphene. Commercialization of this technology is still not progressing due to several challenges – deactivation of bare nanoparticles, use of visible light for the removal of air pollutants using photocatalytic activity, low efficiency of synthesis methods, *etc.* These challenges need to be addressed for the successful implementation of the technology.

REFERENCES

Apte, K., and S. Salvi. 2016. Household air pollution and its effects on health [version 1; peer review: 2 approved]. *F1000Research* 5 (2593):F1000.

Bojdys, Michael J., Jens-Oliver Müller, Markus Antonietti, and Arne Thomas. 2008. Ionothermal synthesis of crystalline, condensed, graphitic carbon nitride. *Chemistry* 14 (27):8177–8182.

Cao, Jun-ji, Yu Huang, and Qian Zhang. 2021. Ambient air purification by nanotechnologies: from theory to application. *Catalysts* 11 (11):1276.

Chaudhary, Abhishek, and Stefanie Hellweg. 2014. Including indoor offgassed emissions in the life cycle inventories of wood products. *Environmental Science & Technology* 48 (24):14607–14614.

Chin, J.-Y., C. Godwin, E. Parker, et al. 2014. Levels and sources of volatile organic compounds in homes of children with asthma. *Indoor Air* 24 (4):403–415.

Dong, Fan, Meiya Ou, Yanke Jiang, Sen Guo, and Zhongbiao Wu. 2014. Efficient and durable visible light photocatalytic performance of porous carbon nitride nanosheets for air purification. *Industrial & Engineering Chemistry Research* 53 (6):2318–2330.

Dong, Fan, Yanjuan Sun, Liwen Wu, Min Fu, and Zhongbiao Wu. 2012. Facile transformation of low cost thiourea into nitrogen-rich graphitic carbon nitride nanocatalyst with high visible light photocatalytic performance. *Catalysis Science & Technology* 2 (7):1332–1335.

Dong, Fan, Zhenyu Wang, Yanjuan Sun, Wing-Kei Ho, and Haidong Zhang. 2013. Engineering the nanoarchitecture and texture of polymeric carbon nitride semiconductor for enhanced visible light photocatalytic activity. *Journal of Colloid and Interface Science* 401:70–79.

Dong, Fan, Liwen Wu, Yanjuan Sun, Min Fu, Zhongbiao Wu, and S. C. Lee. 2011. Efficient synthesis of polymeric g-C$_3$N$_4$ layered materials as novel efficient visible light driven photocatalysts. *Journal of Materials Chemistry* 21 (39):15171–15174.

Le Bechec, M., N. Kinadjian, D. Ollis, R. Backov, and S. Lacombe. 2015. Comparison of kinetics of acetone, heptane and toluene photocatalytic mineralization over TiO2 microfibers and Quartzel® mats. *Applied Catalysis B: Environmental* 179:78–87.

Pui, David Y. H., Sheng-Chieh Chen, and Zhili Zuo. 2014. PM2.5 in China: measurements, sources, visibility and health effects, and mitigation. *Particuology* 13:1–26.

Rao, Zepeng, Xiaofeng Xie, Xiao Wang, et al. 2019. Defect chemistry of Er3+-doped TiO$_2$ and its photocatalytic activity for the degradation of flowing gas-phase VOCs. *The Journal of Physical Chemistry C* 123 (19):12321–12334.

Wang, Yong, Xinchen Wang, and Markus Antonietti. 2012. Polymeric graphitic carbon nitride as a heterogeneous organocatalyst: from photochemistry to multipurpose catalysis to sustainable chemistry. *Angewandte Chemie International Edition in English* 51 (1):68–89.

Weon, Seunghyun, and Wonyong Choi. 2016. TiO$_2$ nanotubes with open channels as deactivation-resistant photocatalyst for the degradation of volatile organic compounds. *Environmental Science & Technology* 50 (5):2556–2563.

Weon, Seunghyun, Jungwon Kim, and Wonyong Choi. 2018. Dual-components modified TiO$_2$ with Pt and fluoride as deactivation-resistant photocatalyst for the degradation of volatile organic compound. *Applied Catalysis B: Environmental* 220:1–8.

Xiang, Quanjun, Jiaguo Yu, and Mietek Jaroniec. 2011. Preparation and enhanced visible-light photocatalytic H$_2$-production activity of graphene/C$_3$N$_4$ composites. *The Journal of Physical Chemistry C* 115 (15):7355–7363.

Yan, S. C., Z. S. Li, and Z. G. Zou. 2009. Photodegradation performance of g-C$_3$N$_4$ fabricated by directly heating melamine. *Langmuir* 25 (17):10397–10401.

Zahra, Rafiei-Sarmazdeh, Zahedi-Dizaji Seyed Morteza, and Kang Aniseh Kafi. 2019. Two-dimensional nanomaterials. In *Nanostructures*, edited by A. Sadia, M. S. Akhtar, and S. Hyung-Shik. Rijeka: IntechOpen.

Zhang, Jinshui, Xiufang Chen, Kazuhiro Takanabe, et al. 2010. Synthesis of a carbon nitride structure for visible-light catalysis by copolymerization. *Angewandte Chemie International Edition in English* 49 (2):441–444.

Zhang, Jinshui, Jianhua Sun, Kazuhiko Maeda, et al. 2011. Sulfur-mediated synthesis of carbon nitride: band-gap engineering and improved functions for photocatalysis. *Energy & Environmental Science* 4 (3):675–678.

Zhang, Yanhui, Zi-Rong Tang, Xianzhi Fu, and Yi-Jun Xu. 2011. Nanocomposite of Ag–AgBr–TiO$_2$ as a photoactive and durable catalyst for degradation of volatile organic compounds in the gas phase. *Applied Catalysis B: Environmental* 106 (3):445–452.

Zheng, Yao, Jian Liu, Ji Liang, Mietek Jaroniec, and Shi Zhang Qiao. 2012. Graphitic carbon nitride materials: controllable synthesis and applications in fuel cells and photocatalysis. *Energy & Environmental Science* 5 (5):6717–6731.

Zhu, Xiaobing, Can Jin, Xiao-Song Li, et al. 2017. Photocatalytic formaldehyde oxidation over plasmonic Au/TiO$_2$ under visible light: moisture indispensability and light enhancement. *ACS Catalysis* 7 (10):6514–6524.

7 2D Nanomaterial-Based Photocatalytic Membranes for Water Treatment

Special Focus on the Scope and Challenges of g-C₃N₄

Merin Joseph, M. J. Avani, Sebastian Nybin Remello, Sibu C. Padmanabhan, Michael A. Morris and Suja Haridas

7.1 INTRODUCTION

Safe potable water is a prime requisite for the sustenance of life on earth. The rapid growth in the industrial sector has led to an inevitable deterioration of water quality necessitating urgent remedial action. Water treatment technologies are in a phase of continuous evolution to attain maximal efficiency (Gupta et al. 2012; Rout et al. 2021; Andreozzi et al. 1999; Yin and Baolin 2015). Among the several available strategies, membrane separation technology offers the advantages of operational simplicity, low energy consumption, excellent separation efficiency, and minimal environmental impacts. The main challenge faced in this regard is the fouling of the membranes due to organic matter deposition resulting in reduced water flux and poor sustainability of efficiency. Photocatalytic water treatment offers a green technology involving the usage of inexhaustible solar energy for the elimination of organic pollutants without the generation of secondary waste (Chong et al. 2010; Doll and Frimmel 2005). The major hurdle faced in this context is the post-treatment separation of the photocatalytic material, especially due to nano dimensions. In the current scenario of water treatment technologies, photocatalytic membranes (PMs) are securing extensive attention as they integrate conventional membrane filtration with photocatalytic degradation in a single unit. The PMs, alongside their filtration capability, degrade the pollutants in the feed solutions *via* the generation of reactive oxygen species (ROS) upon light irradiation. The subsequent lowering of the pollutant concentration aids in preventing the formation of an organic layer on the membrane surface thereby

DOI: 10.1201/9781003343899-7

vividly reducing pore blockage and fouling (Leong et al. 2014; Nasir et al. 2021; Shi et al. 2019a). This significantly reduces the pollutant concentration in the retentate thereby improving the permeate quality. The photocatalytic antibacterial strategy also affords a safe route for bacterial disinfection perhaps due to the penetration and subsequent cell inactivation. The PMs can thus outpace conventional membranes in terms of improved antifouling properties and permeate quality.

After an initial phase of UV-responsive PMs, the trend has now drifted to visible-light-responsive PMs (Shi et al. 2019a). The crucial element in sustainable membrane technology is the right choice of materials for membrane fabrication. Common materials in use include polymers and ceramic materials, each having its recompenses and shortcomings. While cost-effectiveness, abundant resources, and tailorable physical properties render polymeric material attractive, ceramic materials are particularly sought for their chemical stability, mechanical strength, and fouling resistance. Despite tremendous progress in this field both polymeric and ceramic materials still suffer from inherent shortages for industrial applications (Cui et al. 2021c). The pursuit for improvisation and design of new hybrid materials is still ongoing. Developing a highly competent PM can not only overcome the disadvantages of conventional membranes but also can immobilize the photocatalysts to enhance their activity and alleviate the traditional workup procedure indispensable for photocatalytic treatment (Leong et al. 2014; Nasir et al. 2021; Shi et al. 2019b). 2D materials with visible light sensitivity are of vital implications in the design of self-cleaning PMs and have demonstrated exceptional performance in water treatment (Dervin et al. 2016; Hegab and Linda 2015).

7.2 2D MATERIAL-BASED PMs

TiO_2 has been the most-ever explored photocatalytic material on account of its high efficiency, chemical stability, and nontoxicity (Guo et al. 2019; Peiris et al. 2021). The prime shortcoming of the material includes water insolubility and zero visible light sensitivity. Even though TiO_2 nanowires show high water dispersibility, their efficiency is still limited by visible light insensitivity and high electron-hole recombination rate. Graphene oxide (GO) is yet another two-dimensional material with abundant carboxyl, hydroxyl, carbonyl, and other functionalities imparting hydrophilicity (Feng et al. 2020). The negatively charged GO nanosheets upon sonication get completely dispersed in water forming nanochannels. The water purification action is thus a consequence of electrostatic repulsion and molecular sieving. The limitations of GO membranes include low flux and poor mechanical properties. In general, graphene-based films can be obtained either by the assembly of the reduced GO (rGO) sheets or by the reduction of stacked GO films. The removal of surface functional groups and reconstruction of π bonds reduces the hydrophilicity and narrow downs the channels hampering the potential application as filtration membranes (Guo et al. 2019; Nair et al. 2012). Assembly of rGO sheets thus constitutes a prospective strategy for the construction of GO membranes. Chemical reduction of GO has been tried to enhance water flux (Szabó et al. 2005). Intercalation/decoration of GO sheets with nanomaterials can facilitate the assembly of GO into stacked membranes. The nanomaterials act as support effectively preventing the collapse of

GO-stacked plane channels and improving the permeation flux (Han et al. 2015; Xu Chao and Xin Wang 2009; Du et al. 2011; Geng et al. 2010; Li et al. 2010, 2011; Zhou et al. 2010; Pastrana-Martinez et al. 2012, 2015; Xu et al. 2014). *In situ* reduction in this case retains the planar channels to a greater extent (Zhang et al. 2019). The use of polydopamine as a reducing agent to partially reduce GO or for facilitating the binding of different moieties has been reported. In comparison with TiO_2 nanoparticles, nanowires and nanorods have been found to have a better performance when integrated with GO (Liu et al. 2017, 2020; Nair and JagadeeshBabu 2017). TiO_2 nanowires modified by dopamine (PDA) and intercalated into the GO layer significantly improved the water flux and cleansing effect of the membrane (Yu et al. 2020). Integration of 2D GO sheets, 1D nanowires, and Ag nanoparticles synthesized by a two-phase route exhibited enhanced activity in the degradation of organic dyes and phenol (Liu et al. 2018). A facile layer-by-layer synthetic approach to deposit TiO_2 and GO on a polysulfone (PSf) base membrane enabled photoactivity under both UV and sunlight (Gao et al. 2014). Interconnecting TiO_2 nanowires with Fe_2O_3 nanoparticles and GO sheets and fabricated to a membrane by hydrothermal and colloidal blending followed by vacuum filtration significantly showed enhanced humic acid removal from water (Rao et al. 2016).

Layered double hydroxides (LDHs) are 2D semiconducting materials with a positively charged matrix layer and negatively charged intercalated ions. Owing to the advantages like low cost, compositional variability, high chemical stability, exchangeability of interlayer anions, and uniformity in the distribution of metal cations, LDHs and their modifications with other materials have found applications in many fields including photocatalysis (Song et al. 2019; Mohapatra et al. 2016). Inserting LDH between GO layers leads to an enhancement in the interlayer spacing. It also provides more channels and enhances the water flux and purification ability (Liu et al. 2021b). Heterojunction formation between LDH and GO is reported to be facilitated by polydopamine (Wang et al. 2022b). Under the irradiation of visible light, the photo-excited electrons in the conduction band of the TiO_2 nanowire will directly recombine with the holes on the valence band of LDH, thereby creating a hole-rich valence band on TiO_2 and electron-rich conduction band on LDH. The electrons undergo easy transfer to rGO *via* PDA. The efficient charge separation contributes to the enhanced photocatalytic performance of the membrane.

Metal Organic Frameworks (MOFs) are porous materials containing metal ions as nodes linked to organic moieties forming cage-like structures. Large surface area coupled with controllable pore structure and abundant functional groups enables diverse applications. $UiO-66-NH_2$ is a water-stabilized MOF with high affinity to amino functionalities and amenable to integration into the GO matrix. $Ag_2CO_3@UiO-66-NH_2$ (AgCO@UiO) incorporated as the interlayer material between GO sheets proved efficient for the removal of dyes and Cr(VI) (Zeng et al. 2020). The intercalation of Fe-based MOFs into GO nanosheets is reported to regulate 2D nanochannels along with endowing the membrane with photo-Fenton catalytic activity (Li et al. 2021c; Xie et al. 2020).

3D Zr-porphyrin MOF with excellent photocatalytic activity and uniform 1D open triangular channels embedded in a polyamide active layer lends visible light photocatalytic ability along with high membrane permeate flux due to the generation of additional transport channels (Zhao et al. 2021). By electrospinning deacetylated cellulose acetate/polyvinyl pyrrolidone (CeP) nanofibers as cores and beta hydroxyl oxidize iron decorated iron-based MOF (β-FeOOH@MIL-100(Fe)) as photocatalytic sheaths a self-cleaning photocatalytic electrospun nanofiber membrane capable of delivering robust separation efficiency and membrane reusability even for a complex wastewater system was devised (Lu et al. 2021). The promoting effect of Ag@AgCl on the performance of MIL(100)Fe supported on flexible carboxymethylated cellulose fiber *via* surface plasmon resonance (SPR) and its enhanced performance in complex wastewater treatment was demonstrated by Lu et al. (2020). Carboxymethylation significantly improves the hydrophilicity of cotton fabric and the deposition amount of MIL-100(Fe) and could achieve a removal efficiency above 99% for both organic dyes and oil-water emulsions. H_2N-MIL-125 MOFs deposited on $PVA/PAA/SiO_2$ fibrous membranes exhibited high absorption capacities for dyes; the adsorption following Langmuir isotherm and pseudo-second-order kinetics (Huang et al. 2021a). MOF-based PM (CdS/MIL-101) also finds successful application in membrane bioreactors to mitigate biofouling (Ni et al. 2021). Analysis of the biofilm on the fouled membranes revealed degradation of the organic foulants (especially extracellular polymeric substances) and effective inactivation of live bacteria under visible irradiation. A stable electrospun nanofibrous membrane fabricated by *in situ* growing $H_3PW_{12}O_{40}$ @UiO-66 crystals onto crosslinked polyacrylic acid (PAA)-poly(vinyl alcohol) (PVA) nanofibers exhibited photocatalytic degradation of methyl orange and formaldehyde in aqueous solution (Li et al. 2021a). While a reductive pathway was proposed for the degradation of methyl orange, formaldehyde was oxidized. Similarly, covalent organic frameworks grown over carbon nanotubes (CNTs) provide excellent matrices for PMs (Xue et al. 2021a).

Incorporating Zn/Al LDH into the polyamide (PA) layer of polyetherimide mixed matrix membrane generated via interfacial polymerization resulted in a hydrophilic nanofiltration membrane capable of rejecting both cationic dyes and inorganic salts (Mutharasi et al. 2020). $ZnWO_4$/NiAl-LDH/polyvinylidene fluoride (PVDF) and NiAlFe-LDH/PDA/PVDF PMs with hierarchical structure and super oleophobicity have been found effective for oil-water separation as well as the degradation of dyes and antibiotics (Zhao et al. 2020a, Zong et al. 2021).

MXene is a class of two-dimensional transition metal carbides or carbonitrides with lamellar structure and nanoscale mass transfer channels due to layer stacking. Tunable interlayer spacing and abundant functional groups contribute to the hydrophilic attribute of the material. The incorporation of MXene nanosheets and N-$Bi_2O_2CO_3$ (BOC) nanoparticles *via* a facile vacuum-filtration process endowed the membrane with high performance due to more permeation channels (Lin et al. 2022). A low oil adhesion and superhydrophilicity were achieved for PVDF membranes with self-assembled two-dimensional MXene nanosheet exhibiting *in situ* mineralization of photocatalyst β-FeOOH on the membrane surface (Hu et al. 2021).

7.3　g-C₃N₄: A METAL-FREE SEMICONDUCTOR PHOTOCATALYST

Graphitic carbon nitride is a 2D metal-free polymeric material consisting of a well-ordered alignment of tri-s-triazine units, the layers being held together by van der Waals interactions. The material was reported long back in 1834 by Berzelius and Liebig, then named as 'melon' (Thomas et al. 2008). In 1922, Franklin outlined the empirical composition of 'melon' to be C_3N_4. Subsequently, Pauling and Sturdivant speculated a tri-s-triazine type structure of C_3N_4 in the year 1937. By 1940, it was established that this material 'melon' has a graphite structure as reported by Redemann and Lucas (Ong et al. 2016). In 1996, Teter and Hemley proposed five structures of C_3N_4, including α-C_3N_4, β-C_3N_4, cubic-C_3N_4, pseudocubic-C_3N_4, and graphitic-C_3N_4 (g-C_3N_4). Subsequently, multiple forms of g-C_3N_4 were reported including s-triazine-based orthorhombic structures, triazine-based hexagonal structures, and heptazine-based structures (tri-s-triazine) (Figure 7.1). However, an enormous focus was conferred on g-C_3N_4 ever since the first report of its photocatalytic activity by Wang et al. (2009). The laminar structure and triangular pores render it a perfect candidate for membrane technology. The g-C_3N_4 is synthesized *via* controlled thermal polymerization and has a narrow bandgap of 2.7 eV. The defect-rich structure contributes further to its visible light sensitivity. The versatility of the material can be attributed to its thermal/chemical stability, nontoxicity, and biocompatibility. Nonetheless, the photocatalytic performance is limited by a high rate of recombination of electron-hole pairs.

The applications of g-C_3N_4 membranes for environmental remediation have been well-reviewed by several groups (Cui et al. 2021a, 2021b, 2021c; Baig et al. 2021; Li et al. 2021b; Zhang et al. 2021a, 2021b, 2021c; Mao, and Han 2022; Wang et al. 2022a, 2022c; Chen et al. 2022a, 2022b, 2022c). Nevertheless, an exclusive review focusing on PMs is still rare. This chapter provides insight into the recent progress in g-C_3N_4-based PMs. Although a synopsis of fabrication techniques is presented, prime focus rests on the diverse applications like dye degradation, removal of organic pollutants and pharmaceuticals, water desalination, CO_2 reduction, and antibacterial activity. The PMs perform the dual role of water filtration and pollutant degradation. The degradation depends on the light-induced generation of ROS (Figure 7.2). The electrons can generate superoxide radicals which eventually get converted to

FIGURE 7.1 tri-s-triazine structures of g-C_3N_4.

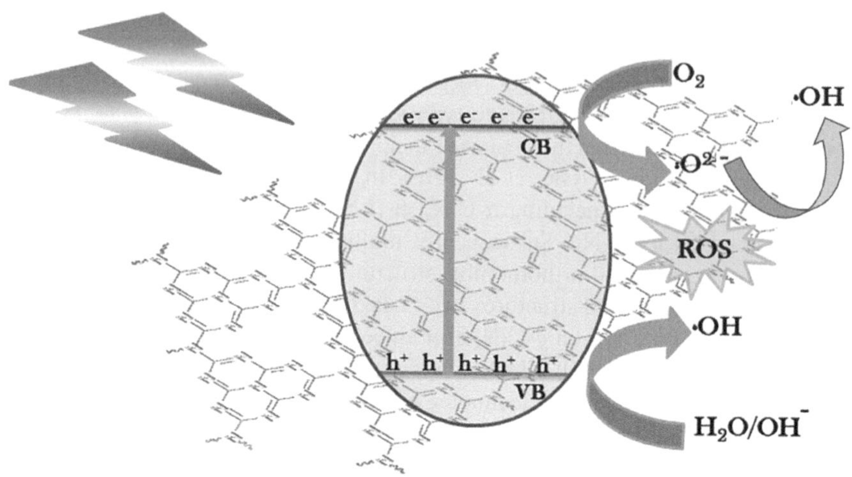

FIGURE 7.2   General scheme of ROS generation.

hydroxyl radicals. The hydroxyl radicals may also be generated at the valence band depending on its relative positioning. The holes can also bring about the oxidation of organic molecules leading to subsequent degradation. A detailed review of the applications of g-C$_3$N$_4$ membranes is provided in the next section. The membranes have been categorized based on their nature to provide a better perception. The challenges and further prospects have also been discussed.

7.4   MEMBRANE FABRICATION STRATEGIES: A GLANCE

g-C$_3$N$_4$ incorporated composite membranes have been explored for their micro, ultra, and nanofiltration performance. Several approaches for the fabrication of g-C$_3$N$_4$-incorporated membranes have been comprehensively covered in the literature (Cui et al. 2021a, 2021b, 2021c; Li et al. 2021b; Zhang et al. 2021a, 2021b, 2021c; Mulungulungu et al. 2022, Mao, and Han 2022; Chen et al. 2022a, 2022b, 2022c; Wang et al. 2020a, 2020b). Even though a detailed discussion of the same is not intended, the section presents an overview of the fabrication strategies. Membrane preparation involves a two-stage process. The first step is the synthesis of single-layer g-C$_3$N$_4$ nanosheets by exfoliation of bulk material to a 2D-laminated structure as represented in Figure 7.3 (Ma et al. 2016; Xu et al. 2013; Lei et al. 2019). Despite the desirable characteristics of a metal-free semiconductor, the performance of g-C$_3$N$_4$ is restricted by limited visible light sensitivity and a high electron-hole recombination rate. Numerous strategies for boosting photocatalytic activity including doping, bandgap engineering, coupling with other semiconductors, dye sensitization, etc. have been sought after (Wang et al. 2022a, 2022c; Zhang et al. 2020, 2021a, 2021b, 2021c). Carbon materials like GO and CNT have been found pertinent as modifiers instigating a significant improvement in hydrophilicity and interlayer spacing leading to high permeation flux (Shi et al. 2019b; Zhao et al. 2016; Qu et al. 2018; Zhang et al. 2018a, 2018b). Functional modification using materials like polydopamine (PDA) is

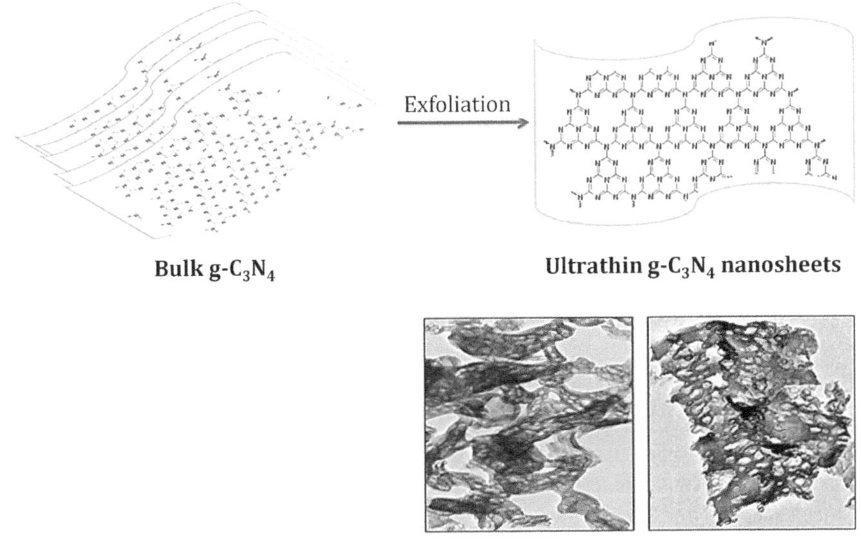

FIGURE 7.3 Schematic illustration of the structural tailoring of g-C_3N_4 by exfoliation.

reported to facilitate the attachment between nanomaterials and also accelerate the rate of photo-induced electron transfer through a two-electron two-proton redox coupling mechanism (Xie et al. 2016; Kim et al. 2014).

The second step involves the integration of g-C_3N_4 nanosheets into the membrane mainly by surface or matrix modification (Figure 7.4). The surface modification strategies include vacuum filtration, dip coating, blending, crosslinking, interfacial polymerization, electrospinning, 3D printing, and grafting induced by ultraviolet (UV)/plasma/chemical agents (Cui et al. 2021a, 2021b, 2021c; Mulungulungu et al. 2022; Chen et al. 2022a, 2022b, 2022c). The key challenge in any of the processes is obtaining a stable dispersion of g-C_3N_4 for solution processing. Vacuum filtration is the most facile and commonly adopted approach. The facile and commonly used method is vacuum or pressure-assisted filtration of the g-C_3N_4 nanosheet suspension on porous support to construct a layer-by-layer deposited membrane (Zhao et al. 2016; Zhou et al. 2016). Hydrogen bonding constitutes the primary binding force between the substrate and the coating layer and is hence susceptible to mechanical damage. Crosslinking with PEG (polyethylene glycol) or glutaraldehyde may prove beneficial in enhancing membrane stability (Huang et al. 2019; Parga et al. 2005; Li et al. 2019b). *In situ* thermal condensation by the immersion of the substrate into the g-C_3N_4 precursor suspension followed by high-temperature calcination yields direct growth of g-C_3N_4 on the membrane (Chang et al. 2018; Park et al. 2016). High-temperature treatment facilitates strong interaction between the species leading to stable coating. The process however demands the critical temperature resistance of the substrate. Chemical grafting offers another synthetic route for membrane modification. Here, inherent functional groups (residual carboxylic/phenolic/amino groups) on the

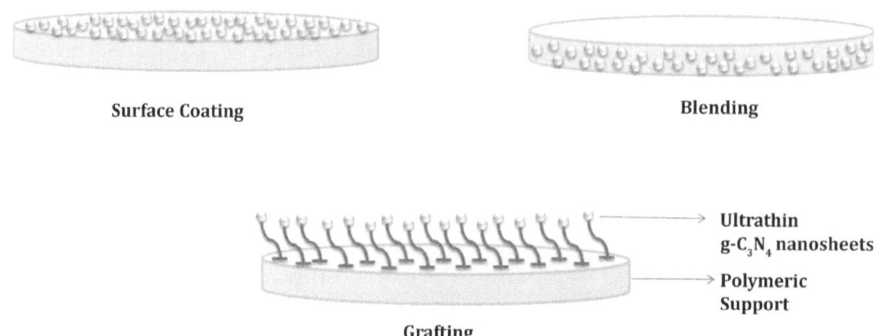

FIGURE 7.4 Common strategies of membrane fabrication.

membrane surface are involved in bond formation with g-C_3N_4. Further facilitation of grafting may be achieved by introducing additional functionalities *via* the chemical treatment of the membranes (Chen et al. 2017; Lee et al. 2018; Chi et al. 2019; Saqib and Aljundi 2016). Precise control of the grafting conditions is quite vital in evading impairment of the membrane pore structure.

The matrix modification involves the one-step synthesis of membranes by blending g-C_3N_4 nanosheets in a polymer solution followed by membrane casting *via* the phase inversion method (Nthunya et al. 2019; Shi et al. 2014). The low mutual diffusivity of the solvent and water induces instability of the casting solution resulting in fast phase separation (Shen et al. 2005). The faster the diffusion of nonsolvent (water) into the casting solution higher will be the extent of pore formation. The process conveniently allows large-scale fabrication of the membrane. The preparation of self-supporting g-C_3N_4 membranes by chemical vapor deposition and vapor deposition polymerization has also been reported (Zhang et al. 2022a). However, these membranes suffer from poor chemical stability and low mechanical strength. Figure 7.3 shows the schematic illustration of three methods for the integration of g-C_3N_4 to membranes such as surface coating, blending, and grafting.

7.5 APPLICATIONS OF PHOTOCATALYTIC g-C_3N_4 MEMBRANES

A schematic representation of the PM filtration is sketched in Figure 7.5. Both polymeric and ceramic membranes have been employed as support materials for the fabrication of g-C_3N_4 membranes. Flexibility and economic aspects favor the usage of polymeric membranes. However, its inherent shortcoming of fouling and vulnerability to UV radiation or ROS generated during light irradiation limits the usage of polymeric supports in PMs. Polytetrafluoroethylene (PTFE) and PVDF are reported to show high stability against UV irradiation (Chin et al. 2006). The resistance has been attributed to the stability of C-F bonds which prevents the disruption of the polymeric structure. PMMA and PDA (polydopamine) have also been found to be resistant to UV radiation, high temperature, and free radical oxidation (Muchtar et al. 2019; Svoboda et al. 2020). Polyethersulfone (PES) is generally characterized by high chemical/thermal stability and mechanical properties. Inorganic materials like

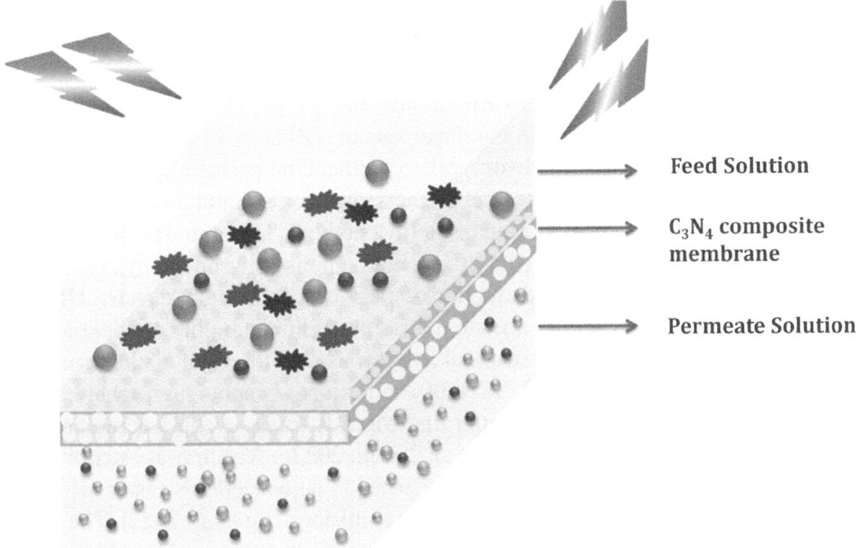

FIGURE 7.5 Schematic representation of PM filtration.

silica and alumina offer better mechanical, thermal, and chemical stability and better resistance to light irradiation preventing physical damage to the membranes. Further, the hydrophilicity imparted due to hydroxyl functionalities can mitigate membrane fouling to some extent. The commercial viability of ceramic membranes is however limited by high production costs. Any feature enhancing the hydrophilicity of the membrane could prove beneficial in the design of an efficient PM.

7.5.1 POLYMER MEMBRANE SUPPORT

g-C_3N_4 nanosheets grown on cellulose-based carbon paper by chemical vapor deposition served as a static photocatalyst and flow-through membrane for the treatment of aqueous methylene blue solution (Dou et al. 2019). The carbon paper could enhance charge transfer maintaining high photocatalytic performance while the oxygen plasma treatment boosted the hydrophilicity. Cellulose nanofibers impregnated with g-C_3N_4 nanosheets *via* simple suction filtration exhibited high water penetration and above 96% photocatalytic efficiency and stability for over 4 hours of continuous operation when tested for dye removal (Zhang et al. 2021a, 2021b, 2021c). g-C_3N_4/GO-based high-performance visible-light-sensitive PM for Rhodamine B and Bisphenol A as model contaminants was assessed by Quan et al. The graphitic carbon nitride nanosheet/rGO composite assembled on commercial cellulose acetate (g-C_3N_4 NS/RGO/CA) by the vacuum filtration and high-pressure process exhibited a pollutant removal efficiency four times higher as compared to simple membrane filtration. In addition, the membranes exhibited good antifouling and bacterial inactivation properties. The *E. coli* removal efficacy may be due to a combined effect of membrane rejection and photocatalytic inactivation. The high flexibility and attractive photocatalytic efficiency could be attributed to the synergistic interaction and linkage between g-C_3N_4 NS and RGO and subsequent facile charge transfer. The use of polydopamine has been reported to simultaneously cause the reduction of GO and also enhance the attachment between GO and g-C_3N_4. The insertion of g-C_3N_4 into GO causes an enlargement in interlayer spacing (Zhao et al. 2016; Li et al. 2017). PDA also serves to enhance the hydrophilicity enhancing permeation flux across the membrane. The PM showed a high efficiency toward a simultaneous separation of oil/water emulsions and degradation of methylene blue. The abundant hydroxyl and amino groups in PDA along with the $\Rightarrow - \Rightarrow$ interactions between methylene blue and g-C_3N_4/GO enable the complete extraction of the dye from the mobile phase. Ultrathin g-C_3N_4 embedded in a cellulose acetate matrix proved to be highly efficient for photocatalytic algal inactivation. Under visible light, 92% of the algal could be removed by the optimal CA/SCN PM (CA/SCN200). Additional benefits highlighted include the elimination of algal organic matter and MC-LR (microcystin-leucine-arginine) *via* the photocatalytic process (Wang et al. 2022a, 2022c). Vacuum-assisted filtration assembling of g-C_3N_4/SiO_2 nanoparticle suspension onto a cellulose ester membrane proved effective for the filtration of oil-water emulsions (Ye et al. 2022). The hydrophilicity of g-C_3N_4 was enhanced by oxygen plasma treatment enabling high dispersion in an aqueous solution. The reduced pore size of the hybrid could be related to the intercalation of SiO_2 nanoparticles into the interlayer of the g-C_3N_4 nanosheets shrinking the cavity structure and pore size. A defect-free interface demonstrated super hydrophilicity and underwater superoleophobicity minimizing pore blockage and causing degradation of the retained pollutants inhibiting membrane fouling. The ultrafiltration membrane retained *E. coli* bacteria through size exclusion and caused their photocatalytic inactivation under visible light irradiation. 1D g-C_3N_4 NT and 2D g-C_3N_4 NS have been used as photocatalytic fillers and integrated into rGO laminar structure and finally assembled on disc-type CA membranes for Rhodamine B removal (Wei et al. 2019). The pre-treatment with polydopamine served to enhance the adhesion of GO to the CA matrix. More wrinkles were created in the crumpled laminar nanostructure of rGO by introducing 1D or 2D g-C_3N_4 into the scaffold and local anisotropic stacking aggregation attributed to enhanced surface roughness. 1D intercalation was found to be more beneficial for the enlarging of rGO laminar structure while 2D insertion led to wrinkles *via* stacking aggregation and increased resistance to water flow. Sub-micron rGO/g-C_3N_4 supported on mixed cellulose ester has also been suggested as a potential desalination membrane (Li et al. 2018). As a metal-free photocatalyst, g-C_3N_4 can high effectively generate electron-hole pairs under UV, which can be exactly captured by the GO flakes ensuring charge separation. UV irradiation partly reduces GO to r-GO depending on intensity and irradiation time. The swelling effect of GO flakes is effectively restrained by photocatalytic reduction, which favors desalination.

PVDF has been identified as a suitable substrate for membrane fabrication on account of its chemical tolerance and mechanical strength and compatibility with functional photocatalytic materials. The phase-inverse method was adopted for the synthesis of g-C_3N_4/Ag_3PO_4-PVDF photocatalytic porous membranes with high permeation flux (Cui et al. 2019). The hybrid membranes exhibited asymmetric porous microstructure on account of the prompt exchange of solvent and nonsolvent during phase inversion and ameliorated by the hydrophilic nature of the g-C_3N_4/Ag_3PO_4/ interface. The hydrophilicity contributed to the electrostatic repulsion of BSA reducing membrane fouling. The generation of superoxide and hydroxyl radicals contributed toward the efficient removal of Rhodamine B. Li et al. reported the fabrication of an Ag/CNQDs/g-C_3N_4/PVDF membrane with excellent activity toward norfloxacin (NOR) degradation (Li et al. 2022). The self-cleaning characteristics were evaluated by using bovine serum albumin as a simulated pollutant. Under neutral conditions, a hydrogen bonding interaction between Ag/CNQDs/g-C_3N_4 and NOR was proposed to enhance the photocatalytic performance and the key role of holes, hydroxyl, and superoxide radicals have been emphasized. Strongly alkaline conditions and acidic conditions impaired the membrane performance. Supramolecular aggregates of melamine, cyanuric acid, and urea in DMSO deposited on PVDF membrane and subsequent crosslinking with glutaraldehyde and polyethylene glycol enabled the synthesis of an MCU-C_3N_4/PVDF membrane tested for its self-cleaning performance and stability for degradation of Rhodamine B and tetracycline in a dead-end filtration set-up (Huang et al. 2019). Integration of GO into the assembly (GO/MCU-C_3N_4/PVDF) greatly enhanced the oil-water separation membrane (Shi et al. 2019b). Apart from the increase in hydrophilicity, an increase in the interlayer spacing of GO could be noticed upon MCU-C_3N_4 intercalation. MCU-C_3N_4 can be easily excited and the electron from the conduction band gets transferred easily to GO ensuring charge separation. g-C_3N_4/PVDF membrane also proved highly stable and recyclable membrane for ultrafiltration of dyes in cross-flow mode (Kolesnyk et al. 2020). Hydroxyl functionalized porous g-C_3N_4 nanosheets immobilized on flexible PVDF substrate *via* sodium alginate binder yielded a flexible and super hydrophilic multifunctional membrane with high flux recovery ratio (FRR) and renewability for the elimination of dyes in static and dynamic mode (Xue et al. 2021b). Further, the rough surface exhibited good underwater superoleophobicity enabling efficient oil-water separation. Liu et al. developed polydopamine-modified layered double hydroxide and graphitic nitrogen carbide (LDH@g-C_3N_4@PDA) which was integrated with GO and suction filtered onto PVDF membrane (Liu et al. 2021b). The unique membrane exhibited excellent hydrophilicity and self-cleaning ability and yielded a separation efficiency of above 90% for a wide range of pollutants-methylene blue, Rhodamine B, gasoline, diesel, and petroleum ether. The excellent properties could be attributed to the high photocatalytic efficiency enabling the generation of ROS as a result of the synergistic interaction between constituents. A high photocatalytic degradation rate of 94.6% for tetracycline was reported over the Z-scheme g-C_3N_4/BiOI system blended into the PVDF matrix and cast by phase inversion. PVDF membrane synergistically enhanced the photocatalytic efficiency and the porous membrane structure assisted the migration of CN/BOI migration to the membrane surface exposing active sites. The photodegradation was assisted by the hydroxyl and superoxide species generated *via* irradiation. The piezoelectric properties of the ®-phase of PVDF were also instrumental in accelerating charge separation (Cui et al. 2021a, 2021b, 2021c). Solvothermal synthesized cesium tungsten oxide/graphitic carbon nitride ($Cs_{0.32}WO_3$@g-C_3N_4) hybrids blended into PVDF and cast into fiber membrane *via* electrospinning demonstrated superior water evaporation and salt rejection efficiency (Tessema et al. 2021). $Cs_{0.32}WO_3$ induces strong NIR absorption and localized SPR while g-C_3N_4 exhibits visible light sensitivity. The photothermal membranes also possessed excellent thermal insulation properties minimizing heat dissipation and thus contributing toward a high water evaporation rate. A minimal salt deposition on the membrane during desalination studies perhaps as a consequence of minimal contact arising from hydrophobicity of the membrane enabling rediffusion of ions into bulk water. This demonstrates the resistance to fouling by seawater ions. Studies on sewage water treatment revealed almost complete removal of dyes, antibiotics, and nitrophenol. Meng et al. proposed a magnetically induced freezing casting method to prepare highly effective macroporous Fe_3O_4/g-C_3N/PVDF membranes (FCMs) (Li et al. 2019a). The FCMs exhibited significantly enhanced visible-light absorption because of the exposure of photocatalytic active sites available on the membrane surface and the macroporous structure facilitating light penetration.

PES membrane modified with silver-modified graphitic carbon nitride (Ag/g-C_3N_4) *via* a phase-inversion method was demonstrated as an ultrafiltration membrane with antifouling performance (Zhang et al. 2017a, 2017b). The addition of Ag/g-C_3N_4 resulted in enlarging of finger-like voids perhaps due to an accelerated demixing process with a consequent reduction in hydraulic resistance contributing to enhanced flux permeate. The enhanced membrane hydrophilicity imparted antifouling capability to the membrane. Entrapping of the photogenerated electrons by Ag nanoparticles prevented the electron-hole recombination facilitating the transfer of charge carriers. The incorporation of silver nanoparticles also contributed to bacterial elimination along with pollutant degradation. Ag_3PO_4 was modified with APTMS as a coupling agent to obtain amino-functionalized nanosilica and incorporated as filler into the PES matrix by phase inversion (Ghalamchi et al. 2019). The nanoparticles also disrupted the polymer chain formation lending porosity and roughness to the membrane. The permeability could be correlated to hydrophilicity and the antibacterial activity correlated both to the active silver ions and photo-assisted generation of free radicals. The high FRR indicated the antifouling characteristics of the membrane. g-C_3N_4-modified MXene incorporated into PES membrane *via* vacuum filtration was proposed as a novel assembly for wastewater treatment (Zeng et al. 2022). MXene possesses the dual advantage of abundant surface functionalities (–OH, –O, etc) enhancing the hydrophilicity and nano-scaled mass transfer channels with easily tailorable layer spacing. Membrane swelling which was identified as the inherent shortcoming of MXenes could be overcome by enhancing the layer spacing through g-C_3N_4 intercalation. Enhanced layer spacing permitted water flux minimizing the swelling effect. The high rejection rate of Congo Red and Trypan Blue was assigned to the electrostatic repulsion of the anionic dyes by the negatively charged membrane surface. Combining MXene with g-C_3N_4 provides an effective and shorter electron transfer channel enhancing electron transfer kinetics which eventually boosts the photocatalytic performance. The degradation of sulphamethoxazole under solar irradiation over carboxylated mesoporous graphitic carbon nitride/TiO_2-modified polysulfone membrane has been investigated (Yu et al. 2018). Oxygen doping of g-C_3N_4 enhanced light sensitivity and suppressed electron-hole recombination leading to high phenol rejection when cast into the PES matrix (Salim et al. 2019). A similar effect was noticed in H_2O_2-C_3N_4/polysulfone composite membrane when tested for the removal of humic acid. The membranes exhibited considerable fouling resistance perhaps due to high hydrophilicity and synergistic enhancement of photocatalytic activity (Salehian et al. 2022). The incorporation polyamide active layer on the PES surface *via* interfacial polymerization reaction between piperazine and trimesoylchloride and subsequent integration of nanomaterials has opted as a potential strategy for the formation of nanofiltration membranes (Chen et al. 2016; Gui et al. 2022). g-C_3N_4 incorporation enhanced surface roughness and hydrophilicity and also imparted negative surface charge density contributing toward antifouling characteristics. Careful optimization of g-C_3N_4 proportion was quite crucial to obtain a significant salt rejection. GO, N-doped graphene (NG), and g-C_3N_4 nanosheets (GO/NG/g-C_3N_4) on PES substrate possessing excellent hydrophilicity, flux permeability, excellent dye rejection performance and the self-cleaning ability for long-term nanofiltration have been reported (Hou et al. 2022).

The incorporation of an active polyamide layer into PES membranes has been adopted as a general strategy to enhance the interaction between support and photoactive material. A PM was fabricated from the polyamide (PA) microfiltration membrane and visible-light-responsive Bi_2WO_6/protonated g-C_3N_4 (BPG) photocatalyst for the elimination of 17β-estradiol (E2), an endocrine-disrupting chemical (Qing et al. 2022). The BPG photocatalyst was embedded in visible light transmitting calcium alginate hydrogels and then immobilized into the PA layer by vacuum filtration and crosslinking photodegradation. Under static and dynamic conditions, the sustained removal efficiency of 95.1% could be achieved mainly due to the synergistic effects of the adsorption of PA and photodegradation of the BPG nanocomposites.

By electrospinning polyacrylonitrile (PAN) and graphite carbon nitride (g-C_3N_4), Jiang et al. synthesized fibrous photocatalytic PAN@g-C_3N_4 air filtration membranes with high efficiency for the removal of formaldehyde (Cui et al. 2021a, 2021b, 2021c). Under simulated sunlight, photogenerated holes could effectively bring about the oxidation of adsorbed HCHO. The addition of g-C_3N_4 resulted in an increase in the surface area roughness and mechanical strength and a reduction in fiber diameter. At an optimal loading, reorientation of the polar groups in g-C_3N_4 during electrospinning induced an increase in charge storage capacity enabling the facile adsorption of charged species and subsequently attaining high filtration efficacy. Trying out the filtration of particulate matter, aggregation on the mesh surface could be visualized from SEM images. The strong surface tension and electrostatic interaction between charged fine particles and fiber filtration media, interception of Brownian motion, and electrostatic effect synergistically contribute toward blocking of pollutants. The self-cleaning and antimicrobial characteristics of C_3N_4-functionalized PAN membranes were also demonstrated by Li et al. (2019b). Protonation of bulk g-C_3N_4 provides high dispersibility and stability and modulates the electronic bandgap for effective charge transfer. A novel composite membrane (PN/Ag) with plasmonic heterojunction was synthesized through the combination of PAN, N-doped carbon dots (NCDs)/g-C_3N_4, and $Ag_2C_2O_4$ by electrospinning technique and successive ionic layer adsorption and reaction process for the efficient photocatalytic disinfection (Xu et al. 2022). The formation of PN/Ag heterojunction enhanced the lifetime of photogenerated charge carriers as established by lifetime measurements and supplemented by Impedance analysis and Nyquist plots. Dispersion of the hydrophilic g-C_3N_4 sheets obtained *via* oxygen plasmonic treatment in dopamine and polyethylenimine and subsequent coating on porous PAN resulted in a hydrophilic high-flux nanofiltration membrane for sustainable treatment of dye and saline water (Ye et al. 2019). Aggregation of PDA through covalent polymerization generated surface roughness and synergistic interaction enabled firm adhesion and robust linkage with g-C_3N_4. The stacking of g-C_3N_4 nanosheets in the bio-inspired PDA/PEI layer created nano-channels for high permeation flux. A 0D/2D heterojunction composite membrane engineered by depositing graphitic carbon nitride nano/microspheres with abundant wrinkles onto the PAA-functionalized CNTs (CNTs-PAA) membrane through hydrogen bonding interactions demonstrated high photocatalytic and antibacterial properties (Ji et al. 2021). In addition to the super hydrophilicity and underwater superhydrophobicity, the formation of J-aggregates endowed a bandgap of 1.77 eV enabling high visible light sensitivity.

Nafion, a perfluorinated-sulfonated polymer, exhibits excellent properties such as good thermal stability, chemical stability, and hydrophilic nanopores and is quite suited for the fabrication of PMs (Zhang et al. 2018a, 2018b). The nano-cavities present in the Nafion/g-C_3N_4 hybrid membranes are perfect to anchor Ag nanocrystals sized 4–5 nm. The Ag nanoparticles along with enhancing charge separation by the formation of the Schottky barrier, promote visible light absorption *via* SPR leading to efficient rejection of Rhodamine B. High visible transmittance and thermal stability of polymethylmethacrylate (PMMA) renders it highly suited for incorporation of photoactive materials. CO_2 reduction in a g-C_3N_4-TiO_2/Nafion PM reactor was proposed as a green pathway for fuel generation utilizing sustainable solar energy in a continuous operating mode (Brunetti et al. 2019). Contact time and feed molar ratio played a primary role in determining reactor performance, allowing the modulation of product removal to circumvent undesired secondary reactions.

g-C_3N_4 immobilized with various amounts of PMMA on polyurethane (PUR) nanofibrous fabric was reported as an efficient PM for the treatment of methylene blue (Svoboda et al. 2020). When the fabric was immersed in the PMMA solution, thin layers of PMMA were created around the PUR nanofibers and the fibers were connected *via* PMMA bridges as observed from SEM images. The exfoliated C_3N_4 layer could be easily fixed to PUR through dip coating, the amount of loading is dependent on the PMMA content. The high sorption of MB was attributed to the electrostatic interactions between cationic dye and negatively charged membrane. H_2O_2-assisted membrane regeneration was opted for as the photocatalytic performance was dependent on superoxide radicals owing to the specific positioning of VB and CB. Polyvinylpyrrolidone (PVP)-coated porous potassium doped g-C_3N_4 (PKCN) membrane fabricated by vacuum filtration displayed high photon-Fenton self-cleaning ability, high FRR for oil-water separation (Yue et al. 2022). An iron-based 3-D MOF material (MIL-88A) blended with g-C_3N_4 in PAN matrix in conjunction with poly-sulfone/MoS_2 and cast into chitosan matrix with high photo-Fenton performance was reported by Zheng et al. (2021). The synergistic effect of the PAN@MIL-88A/g-C_3N_4 catalytic layer and PSF(polysulfone)@MoS_2 promoter layer effectively generated hydroxyl radicals to degrade methylene blue imparting antifouling characteristics to the membrane. To improve the interfacial affinity and antifouling properties of the polyphenyl sulfone (PPSU) membrane, nano CuO/g-C_3N_4 sheets were synthesized *via* one-pot facile calcination and integrated into the PPSU matrix *via* phase inversion

(Arumugham et al. 2019). Under optimized conditions, the ultrafiltration membrane displayed a high protein rejection rate while maintaining a remarkable permeation flux and FRR. GO/g-C$_3$N$_4$/Ag composite membrane was designed and fabricated by a simple filtration-assisted assembly over polycarbonate support (Zhao et al. 2020b). The intercalated Ag NPs and g-C$_3$N$_4$ facilitate the formation of hot spots for SERS detection and active sites for photocatalytic degradation of paraoxon-ethyl thus yielding a multifunctional membrane. Paraoxon-ethyl at nM-level concentrations could be detectable and then degraded almost completely under visible-light irradiation. The high SERS activity was ascribed to the integration of the enrichment ability of GO and SPR of Ag NPs while the degradation was attributed to the efficient generation of active radicals and sustained activity due to charge separation.

PTFE finds application in air and water purification due to its good chemical stability, high strength, low friction, corrosion resistance, low surface energy, and excellent hydrophobicity. S-C$_3$N$_4$/PTFE membrane displayed high prospects for sewage water treatment (Chen et al. 2022b). High rejection of tetracycline, humic acid, and tryptophan-like compounds could be visualized, the efficiency being a synergistic effect of morphological characteristics and enhanced photocatalytic performance. Chi et al. (2019) reported the immobilization of g-C$_3$N$_4$-TiO$_2$ on the PTFE ultrafiltration membranes *via* a facile plasma-enhanced surface grafting technique using PAA as a binder. The abundant carboxyl groups in the PAA layer acted as grafting sites to coordinate with the Ti cation. The high performance is related to the super wettability and robust photocatalytic efficiency and a discussion on the different self-cleaning mechanisms under UV and visible light is provided. The Mn$_3$O$_4$/g-C$_3$N$_4$ nanosheets@ PTFE membrane was synthesized by facile vacuum filtration for peroxymonosulfate (PMS) activation to remove 4-chlorophenol (CP) and a series of pollutants (BPA, Rh B, and Levofloxacin) (Chen et al. 2020). The g-C$_3$N$_4$ matrix prevented the excess growth of Mn$_3$O$_4$ leading to the exposure of active sites.

7.5.2 CERAMIC MEMBRANE SUPPORT

The fabrication and microstructures of an inorganic membrane should be carefully considered in the PM systems. The desirable porosity, mechanical strength, pore size distribution could be obtained by controlling the sintering process to achieve high pure water flux and permeability for the hollow fiber membrane. Nanowires/tubes get dispersed easily without aggregation to form membrane materials and the direct contact established with target molecules enhances the photocatalytic performance. Most used ceramic filtration membranes are either silica or alumina-based. The 2D sub-nano channels between the adjacent GO nanosheets can act as molecular sieves, allowing the penetration of small molecules and enabling their use as filtration membranes. Intercalation of nanomaterials enables enlarging of interlayer spacing improving permeation flux. This however contributes toward membrane fouling due to the high adsorptive characteristics of GO for a wide spectrum of pollutants. Activated carbon fiber (ACF), as one of the most promising materials, has attracted much attention due to its high electrical conductivity, flexibility and can be used as a flexible porous support

BiFeO$_3$/g-C$_3$N$_4$ nanorods were coupled to the SiO$_2$ membrane obtained *via* electrospinning of tetraethyl orthosilicate/PVP solution (Hui et al. 2021). The soft spun membrane along with excellent photochemical activity for methyl orange degradation exhibited magnetic recovery, multiple recyclability, and non-secondary pollution due to high surface area. EIS measurement established the photogenerated charge transfer at solid/bulk interfaces. The relative positioning of CB of $BiFeO_3$ and g-C_3N_4 resulted in a type II heterojunction and enabled a facile electron transfer from g-C_3N_4 to $BiFeO_3$. SiO_2/g-C_3N_4/BiOX (X-I, Br) heterojunction membrane was prepared by combining vapor deposition and hydrothermal techniques with electrospinning technology (Yang et al. 2022). BiOX is a typical layered semiconductor of *P4/nmm* space group with strong oxidizing and photocatalytic ability. Morphological analysis revealed BiOX nanoflowers embedded in the pores of the g-C_3N_4 membrane forming a synergistic interface that enabled efficient electron transfer at the heterojunction. The degradation of tetracycline was attributed to the reduction by superoxide radicals accumulated at CB of BiOX and the holes in the VB of g-C_3N_4.

Wang et al. (2017) reported the fabrication of a photoelectrocatalytic g-C_3N_4/CNTs/Al_2O_3 membrane by sequentially coating CNTs layer with high electroconductivity and g-C_3N_4 layer with good visible-light response on Al_2O_3 membrane support. Applying a positive potential to the CNT layer induced migration of photoelectrons creating efficient charge separation. The membrane displayed photo-induced and electro-induced wettability and under optimal voltage supply and visible light irradiation, the phenol rejection peaked at 94%. Phosphorus-doped g-C_3N_4 integrated with Al_2O_3 hollow fiber membrane module was used as a PM reactor for the degradation of methylene blue, methyl orange, phenol, and a mixture of the three (Hu et al. 2019). P was substituted into the C sites and vacancies within heptazine rings of CN units. There was a downshift of the CB upon the incorporation of P into the CN matrix leading to a lower bandgap and enhanced visible light sensitivity. The enhanced photocatalytic activity induces the degradation of pollutants on the membrane surface mitigating membrane fouling.

g-C_3N_4/TiO_2 nanotube array (TNA) membrane was fabricated by immobilization of g-C_3N_4 QDs into a free-standing TNA. Benefiting from the synergistic effect of membrane filtration and photocatalysis, more than 60% of Rhodamine B could be removed from water under visible light irradiation (Zhang et al. 2017a, 2017b). g-C_3N_4/TNA membrane also showed an enhanced anti-fouling ability during filtration of water containing *E. coli* under visible light irradiation, and a permeate flux two times higher than that of filtration alone was obtained by the integrated process. The size of g-C_3N_4 QDs played a crucial role due to the strong quantum confinement and edge effects controlling the optical properties. TNA membranes with straight channels and highly self-ordered arrangements displayed promising performance for water treatment. A reusable membrane composed of g-C_3N_4@TiO_2 nanowires-modified GO (GO/CN@TNWs) was successfully fabricated through electrostatic interactions and a vacuum filtration-assisted self-assembly process (Huang et al. 2021b). The high light capture ability and carrier mobility gained through the introduction of GO and the synergistic interactions between TiO_2 and CN contributed to the high performance of the membranes. The positive zeta potential achieved *via* protonation of CN enabled electrostatic interaction with negatively charged TiO_2 NW forming synergistic heterojunction. The GO negative charges were observed to be partly neutralized by positive charges of CN@TiO_2, the electrostatic interactions once again facilitating the formation of a ternary hybrid. The $\Rightarrow - \Rightarrow$ stacking interactions between GO and organic pollutant molecules contribute to the excellent enrichment efficiency of GO/CN@TNWs membrane (98.7%) toward tetracycline hydrochloride. Due to the negative VB positioning of CN relative to –OH/OH, the photoinduced holes of CN can directly oxidize the target molecules.

Low water flux and heavy membrane fouling hamper the use of graphene-based membranes for water purification and molecular sieving. To circumvent this shortcoming, a mixed dimensional graphene-based membrane, intercalated with Ag nanoparticles in conjunction with g-C_3N_4 (AgNP@g-C_3N_4) as pillars and photocatalysts, was fabricated (Chen et al. 2022a, 2022b, 2022c). The loading of Ag nanoparticles on g-C_3N_4 by photodeposition can greatly improve the photocatalytic ability of g-C_3N_4; meanwhile creating more water transport channels between rGO and g-C_3N_4 laminates to enhance the water permeability. The electrostatic interactions between dye molecules and the membrane were crucial in deciding the rejection ratio. High flux recovery could be obtained even without assistance from H_2O_2 indicating self-cleaning capability. Moreover, a stable water flux and separation performance was maintained by rGO/AgNP@g-C_3N_4 membrane in a cross-flow photocatalytic nanofiltration device.

Bi_2WO_6/g-C_3N_4, a Z scheme photocatalyst integrated into ACF can be used as a membrane photocatalyst to effectively degrade chlortetracycline (CTC) in wastewater (Zhang et al. 2022b). Separation of charges and facile electron transport was confirmed from EIS and PL measurements. The composite membrane could effectively mineralize CTC and the degradation pathway has been sketched. A combination of membrane filtration with catalytic ozonation has been proposed to develop *in situ* self-cleaning of membrane fouling and micropollutants degradation. $CuMn_2O_4$/g-C_3N_4 loaded onto a ceramic membrane was fabricated, where the thickness, roughness, and hydrophilicity/hydrophobicity could be altered by changing the number of coating procedures (Liu et al. 2021a, 2021c). Efficient degradation of tetracycline over g-C_3N_4 assembled on bacterial cellulose and modified with polypyrrole (Ppy@BC/g-C_3N_4) was reported by Zhou et al. (2022). Introduction of oxygen during the active substrate generation stage was proposed to construct a photo electrocatalytic reaction cell. The additional voltage supplied during the process readily supplements the electron transfer thereby reducing the recombination rate.

7.6 PROSPECTS AND CHALLENGES

In short, g-C_3N_4-based PMs have shown substantial capabilities toward various applications such as dye degradation, removal of organic pollutants, removal of pharmaceuticals, water desalination, bacterial inactivation, etc. The process depends on the generation of ROS which brings about the photocatalytic degradation of the organic moieties thereby reducing fouling and enhancing the membrane performance. In general, holes as well as superoxide and hydroxyl radicals are reported to play a major role in deciding the performance of PMs. The choice of support material is also crucial in the design of PMs. The compatibility of the matrix with the photoactive material is highly significant in deciding permeation flux and antifouling characteristics. Polymeric membranes are the most explored support matrices. Both vacuum filtration and phase inversion have been employed. Phase inversion allows precise control of pore characteristics while filtration-assisted deposition enables the formation of a photoactive layer over the support matrix. Specific fabrication strategies like grafting may induce synergistic interaction between support and active species. Increasing the g-C_3N_4 loading may enhance the pollutant removal efficiency, but it may cause a reduction in water flux due to the penetration resistance offered by the g-C_3N_4 layer. In short, there always exists a trade-off between water flux and degradation capability to be optimized during the design of PMs.

g-C_3N_4 is a promising photocatalytic material with unique properties; the limited visible light sensitivity and rapid charge combination are the main challenges to be encountered. The photocatalytic performance can be enhanced *via* typical strategies like cation/anion doping, heterojunction formation, functionalization, etc. The structural and morphological aspects are also highly significant. Exfoliation breaks down the van der Waals interaction enabling the formation of nanosheets. It also facilitates the dispersion of g-C_3N_4 into the solvent during membrane fabrication ensuring the uniform distribution of the material over the membrane surface. The membrane performance is mainly limited by the fouling after continuous treatment. Enhancing hydrophilicity can contribute to a great extent to imparting antifouling characteristics. Oxygen plasma treatment is reported to enhance the hydrophilicity of g-C_3N_4. Integration with carbon materials like GO is highly explored due to a simultaneous enhancement in photocatalytic performance and the enhancement in permeation flux owing to an increase in interlayer spacing and hydrophilicity. The facilitation of electron transfer across the heterojunction boosts the membrane performance. In addition to pollutant removal and oil-water separation, the PMs are also found to affect bacterial elimination. The incorporation of silver nanoparticles seems particularly effective in this concern. Care should be taken to cause effective degradation of the pollutants on the membrane surface. There is a high possibility that small intermediate molecules can pass through the membranes into the permeate reducing water quality. The prospects of such secondary pollution need to be minimized. Further, the scope of photoelectrocatalytic membranes can also be explored since there are only relatively few reports in this regard.

To conclude, despite the continuous improvisations in the field, there exists a huge gap in the practical applications of these membranes. g-C_3N_4-based membrane technology is still in its premature stage. Subsequently, a steadfast approach is quite vital in a comprehensive understanding of the integration of g-C_3N_4 into the membrane matrix. Theoretical backup can assist in procuring a fundamental understanding of the microstructural compatibility and the interfacial properties of the membrane/ g-C_3N_4 system. This will relieve the experimental workload and provide insight into the design of new hybrid materials for next-generation photocatalytic/photoelectrocatalytic membranes.

REFERENCES

Andreozzi, Roberto, Vincenzo Caprio, Amedeo Insola, and Raffaele Marotta. 1999. Advanced oxidation processes (AOP) for water purification and recovery. *Catalysis Today* 53 (1):51–59.

Arumugham, Thanigaivelan, Reshika Gnanamoorthi Amimodu, Noel Jacob Kaleekkal, and Dipak Rana. 2019. Nano CuO/g-C_3N_4 sheets-based ultrafiltration membrane with enhanced interfacial affinity, antifouling and protein separation performances for water treatment application. *Journal of Environmental Sciences* 82:57–69.

Baig, Umair, M. Faizan, and Mohd Sajid. 2021. Semiconducting graphitic carbon nitride integrated membranes for sustainable production of clean water: a review. *Chemosphere* 282:130898.

Brunetti, Adele, Francesca Rita Pomilla, Giuseppe Marcì, Elisa Isabel Garcia-Lopez, Enrica Fontananova, Leonardo Palmisano, and Giuseppe Barbieri. 2019. CO_2 reduction by C_3N_4-TiO_2 Nafion photocatalytic membrane reactor as a promising environmental pathway to solar fuels. *Applied Catalysis B: Environmental* 255:117779.

Chang, Meng-Jie, Wen-Na Cui, Jun Liu, Kang Wang, and Xiao-Jiao Chai. 2018. Fabrication and photocatalytic properties of flexible g-C_3N_4/SiO_2 composite membrane by electrospinning method. *Journal of Materials Science: Materials in Electronics* 29 (8):6771–6778.

Chen, Cheng, Lei Chen, Xiaoying Zhu, and Baoliang Chen. 2022a. Graphene nanofiltration membrane intercalated with AgNP@ g-C_3N_4 for efficient water purification and photocatalytic self-cleaning performance. *Chemical Engineering Journal* 441:136089.

Chen, Congcong, Meng Xie, Lingshuai Kong, Wenhui Lu, Zhenyu Feng, and Jinhua Zhan. 2020. Mn_3O_4 nanodots loaded g-C_3N_4 nanosheets for catalytic membrane degradation of organic contaminants. *Journal of Hazardous Materials* 390:122146.

Chen, Jianxin, Zhiyuan Li, Chongbin Wang, Hong Wu, and Gang Liu. 2016. Synthesis and characterization of g-C_3N_4 nanosheet modified polyamide nanofiltration membranes with good permeation and antifouling properties. *RSC Advances* 6 (113):112148–112157.

Chen, Shuhua, Fang Liu, Rongli Cui, Benjie Zhu, and Xuehui You. 2022b. Removal of tetracycline hydrochloride using S–g-C_3N_4/PTFE membrane under visible light irradiation. *Water Cycle* 3:8–17.

Chen, Wei, Ting Ye, Hang Xu, Taoyuan Chen, Nannan Geng, and Xiaohong Gao. 2017. An ultrafiltration membrane with enhanced photocatalytic performance from grafted N–TiO_2/graphene oxide. *RSC Advances* 7 (16):9880–9887.

Chen, Zhou, Yihong Lan, Yubin Hong, and Weiguang Lan. 2022c. Review of 2D graphitic carbon nitride-based membranes: principles, syntheses, and applications. *ACS Applied Nano Materials* 5 (9):12343–12365.

Chi, Lina, Yingjia Qian, Junqiu Guo, Xinze Wang, Hamidreza Arandiyan, and Zheng Jiang. 2019. Novel g-C_3N_4/TiO_2/PAA/PTFE ultrafiltration membrane enabling enhanced antifouling and exceptional visible-light photocatalytic self-cleaning. *Catalysis Today* 335:527–537.

Chin, Sze Sze, Ken Chiang, and Anthony Gordon Fane. 2006. The stability of polymeric membranes in a TiO_2 photocatalysis process. *Journal of Membrane Science* 275 (1–2):202–211.

Chong, Meng Nan, Bo Jin, Christopher WK Chow, and Chris Saint. 2010. Recent developments in photocatalytic water treatment technology: a review. *Water Research* 44 (10):2997–3027.

Cui, Yahui, Zhenlin Jiang, Chenxue Xu, Min Zhu, Weizhe Li, and Chaosheng Wang. 2021a. Preparation, filtration, and photocatalytic properties of PAN@ gC_3N_4 fibrous membranes by electrospinning. *RSC Advances* 11 (32):19579–19586.

Cui, Yanhua, Lili Yang, Minjia Meng, Qi Zhang, Binrong Li, Yilin Wu, Yunlei Zhang, Jihui Lang, and Chunxiang. 2019. Facile preparation of antifouling g-C_3N_4/Ag_3PO_4 nanocomposite photocatalytic polyvinylidene fluoride membranes for effective removal of rhodamine B. *Korean Journal of Chemical Engineering* 36 (2):236–247.

Cui, Yanhua, Lili Yang, Jian Zheng, Zengkai Wang, Binrong Li, Yan Yan, and Minjia Meng. 2021b. Synergistic interaction of Z-scheme 2D/3D g-C_3N_4/BiOI heterojunction and porous PVDF membrane for greatly improving the photodegradation efficiency of tetracycline. *Journal of Colloid and Interface Science* 586:335–348.

Cui, Yuqi, Xiaoqiang An, Shun Zhang, Qingwen Tang, Huachun Lan, Huijuan Liu, and Jiuhui Qu. 2021c. Emerging graphitic carbon nitride-based membranes for water purification. *Water Research* 200:117207.

Dervin, Saoirse, Dionysios D. Dionysiou, and Suresh C. Pillai. 2016. 2D nanostructures for water purification: graphene and beyond. *Nanoscale* 8 (33):15115–15131.

Doll, T. E., and F. H. Frimmel. 2005. Removal of selected persistent organic pollutants by heterogeneous photocatalysis in water. *Catalysis Today* 101 (3–4):195–202.

Dou, Tianwei, Linlin Zang, Yanhong Zhang, Zhiyao Sun, Liguo Sun, and Cheng Wang. 2019. Hybrid g-C$_3$N$_4$ nanosheet/carbon paper membranes for the photocatalytic degradation of methylene blue. *Materials Letters* 244:151–154.

Du, Jiang, Xiaoyong Lai, Nailiang Yang, Jin Zhai, David Kisailus, Fabing Su, Dan Wang, and Lei Jiang. 2011. Hierarchically ordered macro– mesoporous TiO$_2$– graphene composite films: improved mass transfer, reduced charge recombination, and their enhanced photocatalytic activities. *ACS Nano* 5 (1):590–596.

Feng, Xiaofang, Zongxue Yu, Runxuan Long, Yuxi Sun, Ming Wang, Xiuhui Li, and Guangyong Zeng. 2020. Polydopamine intimate contacted two-dimensional/two-dimensional ultrathin nylon basement membrane supported RGO/PDA/MXene composite material for oil-water separation and dye removal. *Separation and Purification Technology* 247:116945.

Gao, Yong, Meng Hu, and Baoxia Mi. 2014. Membrane surface modification with TiO$_2$–graphene oxide for enhanced photocatalytic performance. *Journal of Membrane Science* 455:349–356.

Geng, Xiumei, Liang Niu, Zhenyuan Xing, Rensheng Song, Guangtong Liu, Mengtao Sun, Guosheng Cheng. 2010. Aqueous-processable noncovalent chemically converted graphene–quantum dot composites for flexible and transparent optoelectronic films. *Advanced Materials* 22 (5):638–642.

Ghalamchi, Leila, Soheil Aber, Vahid Vatanpour, and Mohsen Kian. 2019. A novel antibacterial mixed matrixed PES membrane fabricated from embedding aminated Ag$_3$PO$_4$/g-C$_3$N$_4$ nanocomposite for use in the membrane bioreactor. *Journal of Industrial and Engineering Chemistry* 70:412–426.

Gui, Liangliang, Yuqi Cui, Yuzhang Zhu, Xiaoqiang An, Huachun Lan, and Jian Jin. 2022. g-C$_3$N$_4$ nanofibers network reinforced polyamide nanofiltration membrane for fast desalination. *Separation and Purification Technology* 293:121125.

Guo, Qing, Chuanyao Zhou, Zhibo Ma, and Xueming Yang. 2019. Fundamentals of TiO$_2$ photocatalysis: concepts, mechanisms, and challenges. *Advanced Materials* 31 (50):1901997.

Gupta, Vinod Kumar, Imran Ali, Tawfik A. Saleh, Arunima Nayak, and Shilpi Agarwal. 2012. Chemical treatment technologies for waste-water recycling—an overview. *Rsc Advances* 2 (16): 6380–6388.

Han, Yi, Yanqiu Jiang, and Chao Gao. 2015. High-flux graphene oxide nanofiltration membrane intercalated by carbon nanotubes. *ACS Applied Materials & Interfaces* 7 (15):8147–8155.

Hegab, Hanaa M., and Linda Zou. 2015. Graphene oxide-assisted membranes: fabrication and potential applications in desalination and water purification. *Journal of Membrane Science* 484:95–106.

Hou, Ruitong, Yi He, Hao Yu, Teng He, Yixuan Gao, and Xiao Guo. 2022. A self-cleaning membrane based on NG/g-C$_3$N$_4$ and graphene oxide with enhanced nanofiltration performance. *Journal of Materials Science* 57:1–16.

Hu, Chechia, Mao-Sheng Wang, Chien-Hua Chen, Yi-Rui Chen, Ping-Hsuan Huang, and Kuo-Lun Tung. 2019. Phosphorus-doped g-C$_3$N$_4$ integrated photocatalytic membrane reactor for wastewater treatment. *Journal of Membrane Science* 580:1–11.

Hu, Jiaxin, Yingqing Zhan, Guiyuan Zhang, Qingying Feng, Wei Yang, Yu-Hsuan Chiao, Shirui Zhang, and Ao Sun. 2021. Durable and super-hydrophilic/underwater super-oleophobic two-dimensional MXene composite lamellar membrane with photocatalytic self-cleaning property for efficient oil/water separation in harsh environments. *Journal of Membrane Science* 637:119627.

Huang, Jiming, Ding Huang, Fanbao Zeng, Long Ma, and Zhengbang Wang. 2021a. Photocatalytic MOF fibrous membranes for cyclic adsorption and degradation of dyes. *Journal of Materials Science* 56 (4):3127–3139.

Huang, Jinhui, Jianglin Hu, Yahui Shi, Guangming Zeng, Wenjian Cheng, Hanbo Yu, Yanling Gu, Lixiu Shi, and Kaixin Yi. 2019. Evaluation of self-cleaning and photocatalytic properties of modified g-C$_3$N$_4$ based PVDF membranes driven by visible light. *Journal of Colloid and Interface Science* 541:356–366.

Huang, Yi, Gen Zhu, Kun Zou, Fei Tian, Thakur Prasad Yadav, Hui Xu, Guohai Yang, Haitao Li, and Lulu Qu. 2021b. Highly efficient removal of organic pollutants from wastewater using a recyclable graphene oxide membrane intercalated with g-C$_3$N$_4$@TiO$_2$-nanowires. *Journal of Molecular Liquids* 337:116461.

Hui Zhao, Wen, Shengjuan Ma, Fengzhu Lv, Jiajing Feng, and Yihe Zhang. 2021. Photocatalysis of free-standing electrospinning SiO$_2$ membranes with loaded BiFeO$_3$/C$_3$N$_4$ short rods. *Colloids and Surfaces A: Physicochemical and Engineering Aspects* 628:127326.

Ji, Lingtong, Luke Yan, Min Chao, Mengru Li, Jincui Gu, Miao Lei, Yanmei Zhang, et al. 2021. Sphagnum inspired g-C$_3$N$_4$ nano/microspheres with smaller bandgap in heterojunction membranes for sunlight-driven water purification. *Small* 17 (12):2007122.

Kim, Jae Hong, Minah Lee, and Chan Beum Park. 2014. Polydopamine as a biomimetic electron gate for artificial photosynthesis. *Angewandte Chemie* 126 (25):6482–6486.

Kolesnyk, Iryna, Joanna Kujawa, Halyna Bubela, Viktoriia Konovalova, Anatolii Burban, Aleksandra Cyganiuk, and Wojciech Kujawski. 2020. Photocatalytic properties of PVDF membranes modified with g-C$_3$N$_4$ in the process of Rhodamines decomposition. *Separation and Purification Technology* 250:117231.

Lee, Xin Jiat, Pau Loke Show, Tomohisa Katsuda, Wei-Hsin Chen, and Jo-Shu Chang. 2018. Surface grafting techniques on the improvement of membrane bioreactor: state-of-the-art advances. *Bioresource Technology* 269:489–502.

Lei, Ganchang, Yanning Cao, Wentao Zhao, Zhaojin Dai, Lijuan Shen, Yihong Xiao, and Lilong Jiang. 2019. Exfoliation of graphitic carbon nitride for enhanced oxidative desulfurization: a facile and general strategy. *ACS Sustainable Chemistry & Engineering* 7 (5):4941–4950.

Leong, Sookwan, Amir Razmjou, Kun Wang, Karen Hapgood, Xiwang Zhang, and Huanting Wang. 2014. TiO$_2$ based photocatalytic membranes: a review. *Journal of Membrane Science* 472:167–184.

Li, Bing, Xintong Zhang, Xinghua Li, Lei Wang, Runyuan Han, Bingbing Liu, Weitao Zheng, Xinglin Li, and Yichun Liu. 2010. Photo-assisted preparation and patterning of large-area reduced graphene oxide–TiO$_2$ conductive thin film. *Chemical Communications* 46 (20):3499–3501.

Li, Binrong, Minjia Meng, Yanhua Cui, Yilin Wu, Yunlei Zhang, Hongjun Dong, Zhi Zhu, Yonghai Feng, and Chundu Wu. 2019a. Changing conventional blending photocatalytic membranes (BPMs): focus on improving photocatalytic performance of Fe3O4/g-C$_3$N$_4$/PVDF membranes through magnetically induced freezing casting method. *Chemical Engineering Journal* 365:405–414.

Li, Chen, Tianyi Sun, Guohui Yi, Dashuai Zhang, Yan Zhang, Xiaoxue Lin, Jinrui Liu, Zaifeng Shi, and Qiang Lin. 2022. Fabrication of a Ag/CNQDs/g-C$_3$N$_4$-PVDF photocatalytic composite membrane with excellent photocatalytic and self-cleaning properties. *Journal of Environmental Chemical Engineering* 10 (5):108488.

Li, Fei, Zongxue Yu, Heng Shi, Qiangbin Yang, Qi Chen, Yang Pan, Guangyong Zeng, and Li Yan. 2017. A Mussel-inspired method to fabricate reduced graphene oxide/g-C$_3$N$_4$ composites membranes for catalytic decomposition and oil-in-water emulsion separation. *Chemical Engineering Journal* 322:33–45.

Li, Rui, Yuling Ren, Peixia Zhao, Jing Wang, Jindun Liu, and Yatao Zhang. 2019b. Graphitic carbon nitride (g-C$_3$N$_4$) nanosheets functionalized composite membrane with self-cleaning and antibacterial performance. *Journal of Hazardous Materials* 365:606–614.

Li, Tingting, Zhiming Zhang, Lin Liu, Mingliang Gao, and Zhengbo Han. 2021a. A stable metal-organic framework nanofibrous membrane as photocatalyst for simultaneous removal of methyl orange and formaldehyde from aqueous solution. *Colloids and Surfaces A: Physicochemical and Engineering Aspects* 617:126359.

Li, Xiang, Guohe Huang, Xiujuan Chen, Jing Huang, Mengna Li, Jianan Yin, Ying Liang, Yao Yao, and Yongping Li. 2021b. A review on graphitic carbon nitride (g-C$_3$N$_4$) based hybrid membranes for water and wastewater treatment. *Science of The Total Environment* 792:148462.

Li, Xuyang, Zongxue Yu, Liangyan Shao, Xiaofang Feng, Haojie Zeng, Yuchuan Liu, Runxuan Long, and Ximei Zhu. 2021c. Self-cleaning photocatalytic PVDF membrane loaded with NH2-MIL-88B/CDs and Graphene oxide for MB separation and degradation. *Optical Materials* 119:111368.

Li, Zhangpeng, Jinqing Wang, Xiaohong Liu, Sheng Liu, Junfei Ou, and Shengrong Yang. 2011. Electrostatic layer-by-layer self-assembly multilayer films based on graphene and manganese dioxide sheets as novel electrode materials for supercapacitors. *Journal of Materials Chemistry* 21 (10):3397–3403.

Li, Zhenjiang, Yucheng Xing, Xiaoyan Fan, Liguang Lin, Alan Meng, and Qingdang Li. 2018. rGO/protonated g-C$_3$N$_4$ hybrid membranes fabricated by photocatalytic reduction for the enhanced water desalination. *Desalination* 443:130–136.

Lin, Qingquan, Guangyong Zeng, Guilong Yan, Jianquan Luo, Xiaojie Cheng, Ziyan Zhao, and Han. 2022. Self-cleaning photocatalytic MXene composite membrane for synergistically enhanced water treatment: Oil/water separation and dyes removal. *Chemical Engineering Journal* 427:131668.

Liu, Gonggang, Kai Han, Hongqi Ye, Chenyuan Zhu, Yupei Gao, Yong Liu, and Yonghua Zhou. 2017. Graphene oxide/triethanolamine modified titanate nanowires as photocatalytic membrane for water treatment. *Chemical Engineering Journal* 320:74–80.

Liu, Gonggang, Kai Han, Yonghua Zhou, Hongqi Ye, Xiang Zhang, Jinbo Hu, and Xianjun Li. 2018. Facile synthesis of highly dispersed Ag doped graphene oxide/titanate nanotubes as a visible light photocatalytic membrane for water treatment. *ACS Sustainable Chemistry & Engineering* 6 (5):6256–6263.

Liu, Ye, Zilong Song, Wenhua Wang, ZhenBei Wang, Yuting Zhang, Chao Liu, Yiping Wang, Ao Li, Bingbing Xu, and Fei Qi. 2021a. A CuMn$_2$O$_4$/g-C$_3$N$_4$ catalytic ozonation membrane reactor used for water purification: Membrane fabrication and performance evaluation. *Separation and Purification Technology* 265:118268.

Liu, Yuchuan, Zongxue Yu, Qiuxiang Wang, Ximei Zhu, Runxuan Long, and Xuyang Li. 2021b. Application of sodium dodecyl sulfate intercalated CoAl LDH composite materials (RGO/PDA/SDS-LDH) in membrane separation. *Applied Clay Science* 209:106138.

Liu, Yuchuan, Zongxue Yu, Xiuhui Li, Liangyan Shao, and Haojie Zeng. 2021c. Super hydrophilic composite membrane with photocatalytic degradation and self-cleaning ability based on LDH and g-C$_3$N$_4$. *Journal of Membrane Science* 617:118504.

Liu, Yuchuan, Zongxue Yu, Yixin Peng, Liangyan Shao, Xiuhui Li, and Haojie Zeng. 2020. A novel photocatalytic self-cleaning TiO$_2$ nanorods inserted graphene oxide-based nanofiltration membrane. *Chemical Physics Letters* 749:137424.

Lu, Wanli, Chao Duan, Chaoran Liu, Yanling Zhang, Xin Meng, Lei Dai, Wenliang Wang, Hailong Yu, and Yonghao Ni. 2020. A self-cleaning and photocatalytic cellulose-fiber-supported "Ag@ AgCl@ MOF-cloth" membrane for complex wastewater remediation. *Carbohydrate Polymers* 247:116691.

Lu, Wanli, Chao Duan, Yanling Zhang, Kun Gao, Lei Dai, Mengxia Shen, Wenliang Wang, Jian Wang, and Yonghao Ni. 2021. Cellulose-based electrospun nanofiber membrane with core-sheath structure and robust photocatalytic activity for simultaneous and efficient oil emulsions separation, dye degradation and Cr (VI) reduction. *Carbohydrate Polymers* 258:117676.

Ma, Tengfei, Jie Bai, Haiou Liang, Junzhong Wang, and Chunping Li. 2016. An efficient method for assembling layered g-C_3N_4 nanosheets grow on 1D pore channels carbon fibers as a composite photocatalyst by ultrasound-assisted exfoliation and hydrothermal method. *Vacuum* 134:130–135.

Mohapatra, Lagnamayee, and Kulamani Parida. 2016. A review on the recent progress, challenges and perspective of layered double hydroxides as promising photocatalysts. *Journal of Materials Chemistry A* 4 (28):10744–10766.

Muchtar, Syawaliah, Mukramah Yusuf Wahab, Li-Feng Fang, Sungil Jeon, Saeid Rajabzadeh, Ryosuke Takagi, Sri Mulyati, Nasrul Arahman, Medyan Riza, and Hideto Matsuyama. 2019. Polydopamine-coated poly (vinylidene fluoride) membranes with high ultraviolet resistance and antifouling properties for a photocatalytic membrane reactor. *Journal of Applied Polymer Science* 136 (14):47312.

Mulungulungu, Germain Akonkwa, Tingting Mao, and Kai Han. 2022. Two-dimensional graphitic carbon nitride-based membranes for filtration process: progresses and challenges. *Chemical Engineering Journal* 427:130955.

Mutharasi, Yuvaraj, Noel Jacob Kaleekkal, Thanigaivelan Arumugham, Fawzi Banat, and MSR Sridhar Kapavarapu. 2020. Antifouling and photocatalytic properties of 2-D Zn/Al layered double hydroxide tailored low-pressure membranes. *Chemical Engineering and Processing-Process Intensification* 158:108191.

Nair, Abhinav K., and P. E. JagadeeshBabu. 2017. TiO_2 nanosheet-graphene oxide based photocatalytic hierarchical membrane for water purification. *Surface and Coatings Technology* 320:259–262.

Nair, R. R., H. A. Wu, Parthipan N. Jayaram, Irina V. Grigorieva, and A. K. Geim. 2012. Unimpeded permeation of water through helium-leak–tight graphene-based membranes. *Science* 335 (6067):442–444.

Nasir, Atikah Mohd, Nuha Awang, Juhana Jaafar, Ahmad Fauzi Ismail, Mohd Hafiz Dzarfan Othman, Mukhlis A. Rahman, Farhana Aziz, and Muhamad Azizi Mat Yajid. 2021. Recent progress on fabrication and application of electrospun nanofibrous photocatalytic membranes for wastewater treatment: a review. *Journal of Water Process Engineering* 40:101878.

Ni, Lingfeng, Yijing Zhu, Jie Ma, and Yayi Wang. 2021. Novel strategy for membrane biofouling control in MBR with CdS/MIL-101 modified PVDF membrane by in situ visible light irradiation. *Water Research* 188:116554.

Nthunya, Lebea N., Leonardo Gutierrez, Sebastiaan Derese, Edward N. Nxumalo, Arne R. Verliefde, Bhekie B. Mamba, and Sabelo D. Mhlanga. 2019. A review of nanoparticle-enhanced membrane distillation membranes: membrane synthesis and applications in water treatment. *Journal of Chemical Technology & Biotechnology* 94 (9):2757–2771.

Ong, Wee-Jun, Lling-Lling Tan, Yun Hau Ng, Siek-Ting Yong, and Siang-Piao Chai. 2016. Graphitic carbon nitride (g-C3N4)-based photocatalysts for artificial photosynthesis and environmental remediation: are we a step closer to achieving sustainability? *Chemical Reviews* 116 (12):7159–7329.

Parga, Jose R., David L. Cocke, Jesus L. Valenzuela, Jewel A. Gomes, Mehmet Kesmez, George Irwin, Hector Moreno, and Michael Weir. 2005. Arsenic removal via electrocoagulation from heavy metal contaminated groundwater in La Comarca Lagunera Mexico. *Journal of Hazardous Materials* 124 (1–3):247–254.

Park, Tae Joon, Rajendra C. Pawar, Suhee Kang, and Caroline Sunyong Lee. 2016. Ultra-thin coating g-C_3N_4 on an aligned ZnO nanorod film for rapid charge separation and improved photodegradation performance. *RSC Advances* 6 (92):89944–89952.

Pastrana-Martinez, Luisa M., Sergio Morales-Torres, José L. Figueiredo, Joaquim L. Faria, and Adrián M. T. Silva. (2012). Electrostatic self-assembly of graphene–silver multilayer films and their transmittance and electronic conductivity. *Carbon* 50 (12): 4343–4350.

Pastrana-Martinez, Luisa M., Sergio Morales-Torres, José L. Figueiredo, Joaquim L. Faria, and Adrián M. T. Silva. (2015). Graphene oxide based ultrafiltration membranes for photocatalytic degradation of organic pollutants in salty water. *Water Research* 77:179–190.

Peiris, Sasanka, Haritha B. de Silva, Kumudu N. Ranasinghe, Sanjaya V. Bandara, and Ishanie Rangeeka Perera. 2021. Recent development and future prospects of TiO2 photocatalysis. *Journal of the Chinese Chemical Society* 68 (5):738–769.

Qing, Y., Y. Li, Z. Guo, Y. Yang, and W. Li. 2022. Photocatalytic Bi_2WO_6/g-C_3N_4-embedded in polyamide microfiltration membrane with enhanced performance in synergistic adsorption-photocatalysis of 17β-estradiol from water. *Journal of Environmental Chemical Engineering* 10 (6):108648.

Qu, Lulu, Gen Zhu, Jie Ji, T. P. Yadav, Yijiang Chen, Guohai Yang, Hui Xu, and Haitao Li. 2018. Recyclable visible light-driven g-C_3N_4/graphene oxide/N-carbo nanotube membrane for efficient removal of organic pollutants. *ACS Applied Materials & Interfaces* 10 (49):42427–42435.

Rao, Guiying, Qianyi Zhang, Huilei Zhao, Jiatang Chen, and Ying Li. 2016. Novel titanium dioxide/iron (III) oxide/graphene oxide photocatalytic membrane for enhanced humic acid removal from water. *Chemical Engineering Journal* 302:633–640.

Rout, Prangya R, Tian C. Zhang, Puspendu Bhunia, and Rao Y. Surampalli. 2021. Treatment technologies for emerging contaminants in wastewater treatment plants: a review. *Science of the Total Environment* 753:141990.

Salehian, Saman, Hamid Heydari, Mehran Khansanami, Vahid Vatanpour, and Seyyed Abbas Mousavi. A. 2022. Fabrication and performance of polysulfone/H_2O_2- g-C_3N_4 mixed matrix membrane in a photocatalytic membrane reactor under visible light irradiation for removal of natural organic matter. *Separation and Purification Technology* 285:120291.

Salim, Noor Elyzawerni, N. A. M. Nor, Juhana Jaafar, A. F. Ismail, M. R. Qtaishat, T. Matsuura, M. H. D. Othman, Mukhlis A. Rahman, F. Aziz, and N. Yusof. 2019. Effects of hydrophilic surface macromolecule modifier loading on PES/O g-C_3N_4 hybrid photocatalytic membrane for phenol removal. *Applied Surface Science* 465:180–191.

Saqib, J., and Isam H. Aljundi. 2016. Membrane fouling and modification using surface treatment and layer-by-layer assembly of polyelectrolytes: state-of-the-art review. *Journal of Water Process Engineering* 11:68–87.

Shen, Fei, Xiaofeng Lu, Xiaokai Bian, and Liuqing Shi. 2005. Preparation and hydrophilicity study of poly (vinyl butyral)-based ultrafiltration membranes. *Journal of Membrane Science* 265 (1–2):74–84.

Shi, Yahui, Jinhui Huang, Guangming Zeng, Wenjian Cheng, and Jianglin Hu. 2019a. Photocatalytic membrane in water purification: is it stepping closer to be driven by visible light? *Journal of Membrane Science* 584:364–392.

Shi, Yahui, Jinhui Huang, Guangming Zeng, Wenjian Cheng, Jianglin Hu, Lixiu Shi, and Kaixin Yi. 2019b. Evaluation of self-cleaning performance of the modified g-C_3N_4 and GO based PVDF membrane toward oil-in-water separation under visible-light. *Chemosphere* 230:40–50.

Shi, Yongqian, Saihua Jiang, Keqing Zhou, Chenlu Bao, Bin Yu, Xiaodong Qian, Bibo Wang, et al. 2014. Influence of g-C_3N_4 nanosheets on thermal stability and mechanical properties of biopolymer electrolyte nanocomposite films: a novel investigation. *ACS Applied Materials & Interfaces* 6 (1):429–437.

Song, Biao, Zhuotong Zeng, Guangming Zeng, Jilai Gong, Rong Xiao, Shujing Ye, Ming Chen, Cui Lai, Piao Xu, and Xiang Tang. 2019. Powerful combination of g-C_3N_4 and LDHs for enhanced photocatalytic performance: a review of strategy, synthesis, and applications. *Advances in Colloid and Interface Science* 272:101999.

Svoboda, Ladislav, Nadia Licciardello, Richard Dvorský, Jiří Bednář, Jiří Henych, and Gianaurelio Cuniberti. 2020. Design and performance of novel self-cleaning g-C_3N_4/PMMA/PUR membranes. *Polymers* 12 (4):850.

Szabó, Tamás, Anna Szeri, and Imre Dékány. 2005. Composite graphitic nanolayers prepared by self-assembly between finely dispersed graphite oxide and a cationic polymer. *Carbon* 43 (1):87–94.

Tessema, Aster Aberra, Chang-Mou Wu, Kebena Gebeyehu Motora, and Saba Naseem. 2021. Highly-efficient and salt-resistant CsxWO$_3$@ g-C$_3$N$_4$/PVDF fiber membranes for interfacial water evaporation, desalination, and sewage treatment. *Composites Science and Technology* 211:108865.

Thomas, Arne, Anna Fischer, Frederic Goettmann, Markus Antonietti, Jens-Oliver Müller, Robert Schlögl, and Johan M. Carlsson. 2008. Graphitic carbon nitride materials: variation of structure and morphology and their use as metal-free catalysts. *Journal of Materials Chemistry* 18 (41):4893–4908.

Wang, Dongxu, Juan Chen, Huinan Che, Peifang Wang, and Yanhui Ao. 2022a. Flexible g-C$_3$N$_4$-based photocatalytic membrane for efficient inactivation of harmful algae under visible light irradiation. *Applied Surface Science*, 601:154270.

Wang, Jianlong, and Shizong Wang. 2022. A critical review on graphitic carbon nitride (g-C$_3$N$_4$)-based materials: Preparation, modification and environmental application. *Coordination Chemistry Reviews* 453: 214338.

Wang, Qiuxiang, Zongxue Yu, Yuchuan Liu, Ximei Zhu, Runxuan Long, and Xuyang Li. 2022b. Co-intercalation of TiO$_2$ and LDH to reduce graphene oxide photocatalytic composite membrane for purification of dye wastewater. *Applied Clay Science* 216:106359.

Wang, Shiyu, Liyan Wang, Hongjin Cong, Rui Wang, Jiali Yang, Xinyi Li, Yang Zhao, and Huan Wang. 2022c. A review: g-C$_3$N$_4$ as a new membrane material. *Journal of Environmental Chemical Engineering* 10 (11):108189.

Wang, Siyu, Xiaohui Dai, Fei Li, Jianchao Sun, Chunxia Wu, Yahui Xu, Hai Fan, and Shiyun Ai. 2020a. Floating and stable g-C$_3$N$_4$/PMMA/CFs porous film: an automatic photocatalytic reaction platform for dye water treatment under solar light. *Journal of Porous Materials* 27 (2):465–472.

Wang, Xiaoting, Guanlong Wang, Shuo Chen, Xinfei Fan, Xie Quan, and Hongtao Yu. 2017. Integration of membrane filtration and photoelectrocatalysis on g-C$_3$N$_4$/CNTs/Al$_2$O$_3$ membrane with visible-light response for enhanced water treatment. *Journal of Membrane Science* 541:153–161.

Wang, Xinchen, Kazuhiko Maeda, Arne Thomas, Kazuhiro Takanabe, Gang Xin, Johan M. Carlsson, Kazunari Domen, and Markus Antonietti. 2009. A metal-free polymeric photocatalyst for hydrogen production from water under visible light. *Nature Materials* 8 (1):76–80.

Wang, Yang, Baoyu Gao, Qinyan Yue, and Zhining Wang. 2020b. Graphitic carbon nitride (gC$_3$N$_4$)-based membranes for advanced separation. *Journal of Materials Chemistry A* 8 (37):19133–19155.

Wei, Yibin, Yuxiang Zhu, and Yijiao Jiang. 2019. Photocatalytic self-cleaning carbon nitride nanotube intercalated reduced graphene oxide membranes for enhanced water purification. *Chemical Engineering Journal* 356:915–925.

Xie, Aming, Kun Zhang, Fan Wu, Nana Wang, Yuan Wang, and Mingyang Wang. 2016. Polydopamine nanofilms as visible light-harvesting interfaces for palladium nanocrystal catalyzed coupling reactions. *Catalysis Science & Technology* 6 (6):1764–1771.

Xie, Atian, Jiuyun Cui, Jin Yang, Yangyang Chen, Jihui Lang, Chunxiang Li, Yongsheng Yan, and Jiangdong Dai. 2020. Graphene oxide/Fe (III)-based metal-organic framework membrane for enhanced water purification based on synergistic separation and photo-Fenton processes. *Applied Catalysis B: Environmental* 264:118548.

Xu, Chao, and Xin Wang. 2009. Fabrication of flexible metal-nanoparticle films using graphene oxide sheets as substrates. *Small* 5 (19):2212–2217.

Xu, Chao, Yuelian Xu, and Jiaoli Zhu. 2014. Photocatalytic antifouling graphene oxide-mediated hierarchical filtration membranes with potential applications on water purification. *ACS Applied Materials & Interfaces* 6 (18):16117–16123.

Xu, Jiaxin, Xiuquan Lan, Jianhua Cheng, and Xinhui Zhou. 2022. Facile synthesis of g-C$_3$N$_4$/Ag$_2$C$_2$O$_4$ heterojunction composite membrane with efficient visible light photocatalytic activity for water disinfection. *Chemosphere* 295:133841.

Xu, Jing, Liwu Zhang, Rui Shi, and Yongfa Zhu. 2013. Chemical exfoliation of graphitic carbon nitride for efficient heterogeneous photocatalysis. *Journal of Materials Chemistry A* 1 (46):14766–14772.

Xue, Hongbo, Zhijie Bi, Jiayu Cheng, Sen Xiong, and Yong Wang. 2021a. Coupling covalent organic frameworks and carbon nanotube membranes to design easily reusable photocatalysts for dye degradation. *Industrial & Engineering Chemistry Research* 60 (24):8687–8695.

Xue, Jinjuan, Jiamin Gao, Minjing Xu, Yuqing Zong, Mingxin Wang, and Shuaishuai Ma. 2021b. Super-wetting porous g-C$_3$N$_4$ nanosheets coated PVDF membrane for emulsified oil/water separation and aqueous organic pollutant elimination. *Advanced Materials Interfaces* 8 (19):2100962.

Yang, Jing, Hanyang Song, Ouwen Xu, Shuyu Wan, and Xiashi Zhu. 2022. Preparation of g-C$_3$N$_4$@ bismuth dihalide oxide heterojunction membrane and its visible light catalytic performance. *Applied Surface Science* 583:152462.

Ye, Wenyuan, Hongwei Liu, Fang Lin, Jiuyang Lin, Shuaifei Zhao, Shishi Yang, Jingwei Hou, Shungui Zhou, and Bart Van der Bruggen. 2019. High-flux nanofiltration membranes tailored by bio-inspired co-deposition of hydrophilic gC$_3$N$_4$ nanosheets for enhanced selectivity towards organics and salts. *Environmental Science: Nano* 6 (10):2958–2967.

Ye, Wenyuan, Jinjie Chen, Na Kong, Qingyuan Fang, Mingqiu Hong, Yuxiang Sun, Yifan Li, et al. 2022. Superhydrophilic photocatalytic g-C$_3$N$_4$/SiO$_2$ composite membranes for effective separation of oil-in-water emulsion and bacteria removal. *Separation and Purification Technology* 290:120917.

Yin, Jun, and Baolin Deng. 2015. Polymer-matrix nanocomposite membranes for water treatment. *Journal of Membrane Science* 479:256–275.

Yu, Shuyan, Yining Wang, Faqian Sun, Rong Wang, and Yan Zhou. 2018. Novel mp g-C$_3$N$_4$/TiO2 nanocomposite photocatalytic membrane reactor for sulfamethoxazole photodegradation. *Chemical Engineering Journal* 337:183–192.

Yu, Zongxue, Haojie Zeng, Xia Min, and Xianfeng Zhu. 2020. High-performance composite photocatalytic membrane based on titanium dioxide nanowire/graphene oxide for water treatment. *Journal of Applied Polymer Science* 137 (12):48488.

Yue, Rengyu, and Md Saifur Rahaman. S. 2022. Hydrophilic and underwater superoleophobic porous graphitic carbon nitride (g-C$_3$N$_4$) membranes with photo-Fenton self-cleaning ability for efficient oil/water separation. *Journal of Colloid and Interface Science* 608:1960–1972.

Zeng, Guangyong, Zhenzhen He, Tao Wan, Tairan Wang, Zhaomei Yang, Yongcong Liu, Qingquan Lin, Yiheng Wang, Arijit Sengupta, and Shengyan Pu. 2022. A self-cleaning photocatalytic composite membrane based on g-C$_3$N$_4$ @ MXene nanosheets for the removal of dyes and antibiotics from wastewater. *Separation and Purification Technology* 292:121037.

Zeng, Haojie, Zongxue Yu, Liangyan Shao, Xiuhui Li, Meng Zhu, Yuchuan Liu, Xiaofang Feng, and Ximei Zhu. 2020. Ag$_2$CO$_3$@ UiO-66-NH$_2$ embedding graphene oxide sheets photocatalytic membrane for enhancing the removal performance of Cr (VI) and dyes based on filtration. *Desalination* 491:114558.

Zhang, H., Zhu, Y., Long, J., Ding, Z., Yuan, R., Li, Z. and Xu, C. 2019. In situ construction of layered graphene-based nanofiltration membranes with interlayer photocatalytic purification function and their application for water treatment. *Environmental Science: Nano* 6 (7):2195–2202.

Zhang, Huoli, Jianliang Cao, Peng Kang, Qingjie Tang, Qi Sun, and Mingjie Ma. 2018a. Ag nanocrystals decorated g-C$_3$N$_4$/Nafion hybrid membranes: One-step synthesis and photocatalytic performance. *Materials Letters* 213:218–221.

Zhang, Lilong, Ge Meng, Guifang Fan, Keli Chen, Yulong Wu, and Jian Liu. 2021a. High flux photocatalytic self-cleaning nanosheet g-C$_3$N$_4$ membrane supported by cellulose nanofibers for dye wastewater purification. *Nano Research* 14 (8):2568–2573.

Zhang, Manying, Ziya Liu, Yong Gao, and Li Shu. 2017a. Ag modified g-C$_3$N$_4$ composite entrapped PES UF membrane with visible-light-driven photocatalytic antifouling performance. *RSC Advances* 7 (68):42919–42928.

Zhang, Menglu, Yu Yang, Xiaoqiang An, and Li-an Hou. 2021b. A critical review of g-C$_3$N$_4$-based photocatalytic membrane for water purification. *Chemical Engineering Journal* 412:128663.

Zhang, Qi, Shuo Chen, Xinfei Fan, Haiguang Zhang, Hongtao Yu, and Xie Quan. 2018b. A multifunctional graphene-based nanofiltration membrane under photo-assistance for enhanced water treatment based on layer-by-layer sieving. *Applied Catalysis B: Environmental* 224:204–213.

Zhang, Qi, Xie Quan, Hua Wang, Shuo Chen, Yan Su, and Zhangliang Li. 2017b. Constructing a visible-light-driven photocatalytic membrane by g-C$_3$N$_4$ quantum dots and TiO$_2$ nanotube array for enhanced water treatment. *Scientific Reports* 7 (1):1–7.

Zhang, Rui, Jiacheng Jiang, and Kunlin Zeng. 2022a. Synthesis of Bi$_2$WO$_6$/g-C$_3$N$_4$ heterojunction on activated carbon fiber membrane as a thin-film photocatalyst for treating antibiotic wastewater. *Inorganic Chemistry Communications* 140:109418.

Zhang, Xin, Xingzhong Yuan, Longbo Jiang, Jin Zhang, Hanbo Yu, Hou Wang, and Guangming Zeng 2020. Powerful combination of 2D g-C$_3$N$_4$ and 2D nanomaterials for photocatalysis: recent advances. *Chemical Engineering Journal* 390:124475.

Zhang, Yao, Jingshu Yuan, Yunji Ding, Bo Liu, Liang Zhao, and Shengen Zhang. 2021c. Research progress on g-C$_3$N$_4$ –based photocatalysts for organic pollutants degradation in wastewater: from exciton and carrier perspectives. *Ceramics International* 47 (22):31005–31030.

Zhang, Yizhu, Shangfa Pan, Yuanyuan Zhang, Shaoqiang Su, Xia Zhang, Jian Liu, and Jun Gao. 2022b. Biomimetic high-flux proton pump constructed with asymmetric polymeric carbon nitride membrane. *Nano Research* 16:18–24.

Zhao, Guoqing, Jiao Zou, Xiaoqing Chen, Taiheng Zhang, Jingang Yu, Shu Zhou, Caifeng Li, and Feipeng Jiao. 2020a. Integration of microfiltration and visible-light-driven photocatalysis on a ZnWO$_4$ nanoparticle/nickel–aluminum-layered double hydroxide membrane for enhanced water purification. *Industrial & Engineering Chemistry Research* 59 (14):6479–6487.

Zhao, Huanxin, Shuo Chen, Xie Quan, Hongtao Yu, and Huimin Zhao. 2016. Integration of microfiltration and visible-light-driven photocatalysis on g-C$_3$N$_4$ nanosheet/reduced graphene oxide membrane for enhanced water treatment. *Applied Catalysis B: Environmental* 194:134–140.

Zhao, Lei, Cheng Deng, Sha Xue, Hongbin Liu, Limei Hao, and Mengfu Zhu. 2020b. Multifunctional g-C$_3$N$_4$/Ag NPs intercalated GO composite membrane for SERS detection and photocatalytic degradation of paraoxon-ethyl. *Chemical Engineering Journal* 402:126223.

Zhao, Peixia, Jing Wang, Xinwei Han, Jindun Liu, Yatao Zhang, and Bart Van der Bruggen. 2021. Zr-porphyrin metal–organic framework-based photocatalytic self-cleaning membranes for efficient dye removal. *Industrial & Engineering Chemistry Research* 60 (4):1850–1858.

Zheng, Shengyang, Manhong Huang, Songmei Sun, Haitao Zhao, Lijun Meng, Tianwei Mu, Jialing Song, and Nan Jiang. 2021. Synergistic effect of MIL-88A/g-C$_3$N$_4$ and MoS$_2$ to construct a self-cleaning multifunctional electrospun membrane. *Chemical Engineering Journal* 421:129621.

Zhou, Kai-Ge, Daryl McManus, Eric Prestat, Xing Zhong, Yuyoung Shin, Hao-Li Zhang, Sarah J. Haigh, and Cinzia Casiraghi. 2016. Self-catalytic membrane photo-reactor made of carbon nitride nanosheets. *Journal of Materials Chemistry A* 4 (30):11666–11671.

Zhou, Tang, Lei Zhao, Dingsheng Wu, Quan Feng, and Baobao Zhao. 2022. Uniformly assembled polypyrrole-covered bacterial cellulose/g-C_3N_4 flexible nanofiber membrane for catalytic degradation of tetracycline hydrochloride. *Journal of Water Process Engineering* 47:102775.

Zhou, Yazhou, Juan Yang, Xiaonong Cheng, Nan Zhao, Lei Sun, Hongbo Sun, and Dan Li. 2010. Supraparamagnetic, conductive, and processable multifunctional graphene nanosheets coated with high-density Fe_3O_4 nanoparticles. *ACS Applied Materials & Interfaces* 2 (11):3201–3210.

Zong, Yuqing, Shuaishuai Ma, Jinjuan Xue, Jiandong Gu, and Mingxin Wang. 2021. Bifunctional NiAlFe LDH-coated membrane for oil-in-water emulsion separation and photocatalytic degradation of antibiotic. *Science of the Total Environment* 751:141660.

8 Commercialization of 2D Semiconductors for Environmental Applications

E. J. Jelmy, Rinku Mariam Thomas, C. V. Sijla Rosely and K. P. Anjali

8.1 INTRODUCTION

Semiconducting catalysts have emerged as an effective technique for degrading or removing many types of pollutants. It can decompose a wide range of pollutants into water, carbon dioxide, or other nontoxic intermediates. In photocatalysis, photoinduced electron-hole pair formed in a semiconducting material generates reactive oxygen species (ROS), which then react with the pollutants and degrade them into harmless products. Literature hints that titanium dioxide (TiO_2) is an excellent non-toxic, inexpensive, and stable photocatalyst with a wide bandgap (Padmanabhan et al. 2021). However, the main drawback of TiO_2 system is the limited photocatalytic performance due to the rapid recombination of photogenerated electron-hole pairs and lack of visible light absorption (Anandan et al. 2013). The introduction of two-dimensional (2D) materials such as graphene, reduced graphene oxide (rGO), graphitic carbon nitride, etc., into the conventional catalyst (TiO_2) will enhance the catalytic degradation of the pollutant due to the excellent adsorption properties, good charge separation capacities, fast reaction kinetics, and good light absorption properties achieved by the catalyst due to the association with 2D materials. Besides GO and rGO, MXene-based systems are promising potential semiconductor photocatalysts for catalytic deactivation of pollutants, with favorable properties such as good conductivity, layered structure, mechanical strength, flexibility, large surface area, and high affinity for guest materials. 2D carbides and nitrides (MXenes) with O, – OH, –F, and –Cl groups typically are high surface area systems with porosity, resulting in improved adsorption of guest molecules and contaminants. Another promising candidate for visible-light-enabled photocatalytic inactivation of contaminants is g-C_3N_4, a metal-free photocatalyst. The photocatalytic activity of g-C_3N_4-based photocatalysts is owing to their adequate bandgap of 2.7 eV and valence and conduction band potentials. 2D metal-organic frameworks (MOFs) and covalent organic frameworks (COFs) with high porosity, physicochemical stability, structural tunability, a

DOI: 10.1201/9781003343899-8

wide range of host-guest interactions, sorption, and ion release ability, etc., are class materials with significant photocatalytic activity. Although 2D materials possess many unique properties, such as a very high theoretical specific surface area, high mobility of charge carriers, and good mechanical strength, their commercialized products are still scarce. This chapter briefly reviews the commercial and industrial applications of semiconducting nanomaterials and the importance of incorporating 2D materials in the conventional photocatalyst for property enhancement. The applications covered include wastewater treatment, air purification, hydrogen production, CO_2 reduction, the development of self-cleaning surfaces, and coating.

8.2 RISE OF COMMERCIALIZATION OF 2D SEMICONDUCTORS FOR VARIOUS APPLICATIONS

8.2.1 Water Decontamination

Water is one of the essential components needed for life on earth. The unceasing contamination of water bodies owing to the rapid advancement in the number of industries, agricultural practices, and human lifestyle improvement entails a sustainable solution. Conventional treatment methods have failed in the complete mineralization of organic pollutants, which might result in the formation of other products with high toxicity potential (Sabouni and Gomaa 2019; Enesca 2021). Photocatalysis is a sustainable and environmentally benign treatment method that can attain complete mineralization and zero-end wastes (Sundar and Kanmani 2020), (Zeghioud et al. 2016).

In lab-scale studies, photocatalysts were often applied as suspensions or slurry with smaller volumes of samples and the catalysts were recovered in the subsequent separation stages (Phan et al. 2017). However, the immobilization of photocatalysts using a suitable material would be a better option for practical applications (Han et al. 2021). The key parameters that influence the efficiency of a photocatalytic process are photon transfer and mass transfer. The characteristics of photon transfer and mass transfer vary for suspended and immobilized reactors (Sutisna et al. 2017). In designing a sustainable method for industrial applications, there should be a balance between the process's procedural viability and economic feasibility. Photoreactors can be designed using the catalyst in suspended form or immobilized form. The catalyst in suspended form tends to be more reactive than the immobilized catalyst, but the recovery of the catalyst in suspended reactors is more challenging (Sundar and Kanmani 2020). The immobilized photocatalytic reactors can attain the performance efficiency of the suspended reactors when suitably designed to control the mass transfer and photon transfer limitations by considering the catalyst thickness, bandgap energy, and the distance between the reactant and the catalyst surface (Loeb et al. 2019).

Parabolic trough reactors, flat plate reactors, Heliomans reactors are the commonly used pilot scale solar photoreactors. Besides, double-skin sheet reactors and compound parabolic collector reactors (Ochoa-Gutiérrez et al. 2018) are also gained attention as pilot-scale solar photoreactors. The most widely used commercial solar reactor is the compound parabolic reactor because it combines the capture of maximum solar radiation with suitable hydrodynamics for effective water treatment.

In the design of a reactor setup for environmental applications, different configurations of TiO_2 reactors have been demonstrated, but the pilot and demo plants are still a few. The two essential parameters carefully considered for the scale-up of reactor configurations are degradation efficiency based on reactor output and the ease of catalyst recovery from treated liquid (Abdel-Maksoud, Imam, and Ramadan 2016). In general, immobilized reactors showed better performance characteristics compared to suspended types. However, more studies must be carried out with suspended reactors for improved mass transfer and excellent catalyst recovery methods.

8.2.1.1 Advances in 2D Semiconducting Materials Toward Developing Water Purifiers

The performance of pristine 2D photocatalysts is still unsatisfactory for practical application (Fatima et al. 2022). Surface defect engineering can improve the properties of the 2D photocatalysts by appropriately modifying the local surface microstructure, electronic structure, and carrier concentration (Xiong et al. 2018). Different surface defects types such as anion vacancies, cation vacancies, multiple vacancies, vacancy associates, pits, distortions, and disorders were employed through various strategies, including high-temperature treatment at reducing or inert atmosphere, chemical reduction, vacuum activation, ultraviolet irradiation, phase transformation via fast heating, ball milling, plasma etching, and lithium-induced conversion (Choi, Theerthagiri, and Maia 2021). A mesoporous 2D/2D TiO_2(B)-BiOBr hetero-junction photocatalyst was designed for the treatment of wastewater with multiple contaminants (Han et al. 2021). The coupling structure in the photocatalyst enhances the disconnection of photoexcited carriers and increases the absorption and reaction sites. The hybrid photocatalyst Bi_3TaO_7/Ti_3C_2 displayed superior catalytic performance (99%) in visible-light-driven degradation of methylene blue and showed remarkable stability up to ten cycles of the degradation process (Li et al. 2020).

In the degradation of industrial textile wastewater, the zinc oxide nanoparticles capped with polyethylene glycol as a photocatalyst enhanced the performance of the membrane photocatalytic reactor with high efficiency and less membrane fouling (Desa et al. 2019). Considering the efficiency and cost-effectiveness of polypiperazine-amide (PPA) tight ultrafiltration membrane (UF-PPA) in the reactor unit, it would be an excellent substitute for nanofiltration membranes for dye wastewater treatment. The solar photocatalytic reactor using TiO_2 particles coated onto plastic granules exhibited exceptional efficiency (96.54%) in the degradation of methylene blue solution after 48 h of solar irradiation (Sutisna et al. 2017). The catalyst was positioned at the cavity of the reactor panel made of glass to get maximum efficiency, and the reactor was scaled up by doubling the panel surface area. Consequently, the photodegradation rate increased three-fold with the increase in the catalyst load and reactor volume. Water purification using membranes is always exciting, and the material separation occurs by the screening of particle size and by dissolution-diffusion mechanisms. The significant advantages include low energy consumption, stability, recyclability, inexpensiveness, and facile operational procedures. The research in developing photocatalytic membranes will combine the benefits of membrane technology and photodegradation for pollutant removal. In a study, the membrane of the rGO/polydopamine/$Bi_{12}O_{17}C_{12}$–TiO_2

(rGO/PDA/Bi$_{12}$O$_{17}$C$_{12}$–TiO$_2$) was prepared to remove contaminants effectively (Yu et al. 2019). The prepared nanohybrid dispersion was suction filtered on a cellulose acetate (CA) membrane to develop the proposed purification system (Figure 8.1A). The study hints at the role of rGO in reducing the electron-hole recombination rate and enhancing the photocatalytic activity of the modified TiO$_2$ catalyst (Yu et al. 2019). Similarly, Athanasekou et al. have developed a hybrid photocatalytic/ UF technique for water purification using visible light. The partially rGO/TiO$_2$ composite was prepared, and this was incorporated as a membrane in the water purification device that combines membrane filtration with semiconductor photocatalysis. TiO$_2$-decorated graphene sheets were deposited and stabilized into the pores of UF monochannel monoliths using the dip-coating technique. The filtration experiments were conducted in the dark and with UV light (Figure 8.1B). The results indicate the synergetic effects of graphene oxide on pollutant adsorption and photocatalytic degradation capacity of TiO$_2$ in removing water contaminants (Athanasekou et al. 2014). An efficient graphene oxide/TiO$_2$/bentonite (GO/TiO$_2$/ Bent) sponge was synthesized via an in situ hydrothermal method for water treatment application. The porous, spongy nature of the developed catalysis with a large surface area was found easy to use with minimal loss of TiO$_2$ photocatalyst. Bent with plenty of OH groups enhances the degradation capacity of TiO$_2$ via the formation of photogenerated holes. As in the previous reports, the study hints at the role of rGO in reducing the recombination of charge pairs formed in TiO$_2$ (Liu, Wang, et al. 2020). Combining photoelectrocatalysis and reverse osmosis (RO) offers a viable technique to increase RO performance by eliminating concentrated nitrate in retained flow. For improved photoelectrocatalytic performance, graphene and Ag-loaded TiO$_2$ catalysts were prepared for combined photoelectrocatalysis and RO process (Figure 8.1C). The efficient catalyst was able to reduce 85% of the nitrate primarily into N$_2$ and NO$_2$ and the remaining into NO$_2^-$ and NH$_4^+$ ions at optimized conditions. Furthermore, the photoelectrocatalytic process based on graphene/Ag/N-TiO$_2$ catalyst was coupled with RO to eliminate nitrate in concentrated retentate of the RO via reclamation (Sheydaei and Ayoubi-Feiz 2021). 2D MXene has attracted particular attention, and a couple of investigations have been reported to develop membranes for water treatment. The etching process during the preparation of MXene from the MAX phase leads to abundant hydrophilic functional groups, which can accelerate the hydrophilicity and fouling resistance of polymer membranes. To demonstrate this, Huang et al. have prepared a polymer composite membrane with TiO$_2$@ MXene as the modifier and active material. Polyethersulfone-TiO$_2$@MXene composite membrane developed via a one-step hydrothermal oxidation method (Figure 8.1D). The filtration capacities and the self-cleaning abilities of the composite membrane were analyzed, and this hints at the feasibility of the materials for practical applications (Huang et al. 2021). In a study, TiO$_2$-rGO nanocomposites were used as the photocatalysts in the solar-assisted pilot plant scale for the photodegradation of a mixture of persistent and bio-recalcitrant pollutants (methomyl, pyrimethanil, isoproturon, and alachlor, all used as pesticides) in deionized water (Figure 8.1E). The study aims to remove the pesticides successfully, and at very low concentrations of pesticide, the complete removal was achieved by the catalyst within 25 min, showing the efficiency of the synthesized TiO$_2$-rGO nanocomposites (Luna-Sanguino et al. 2020).

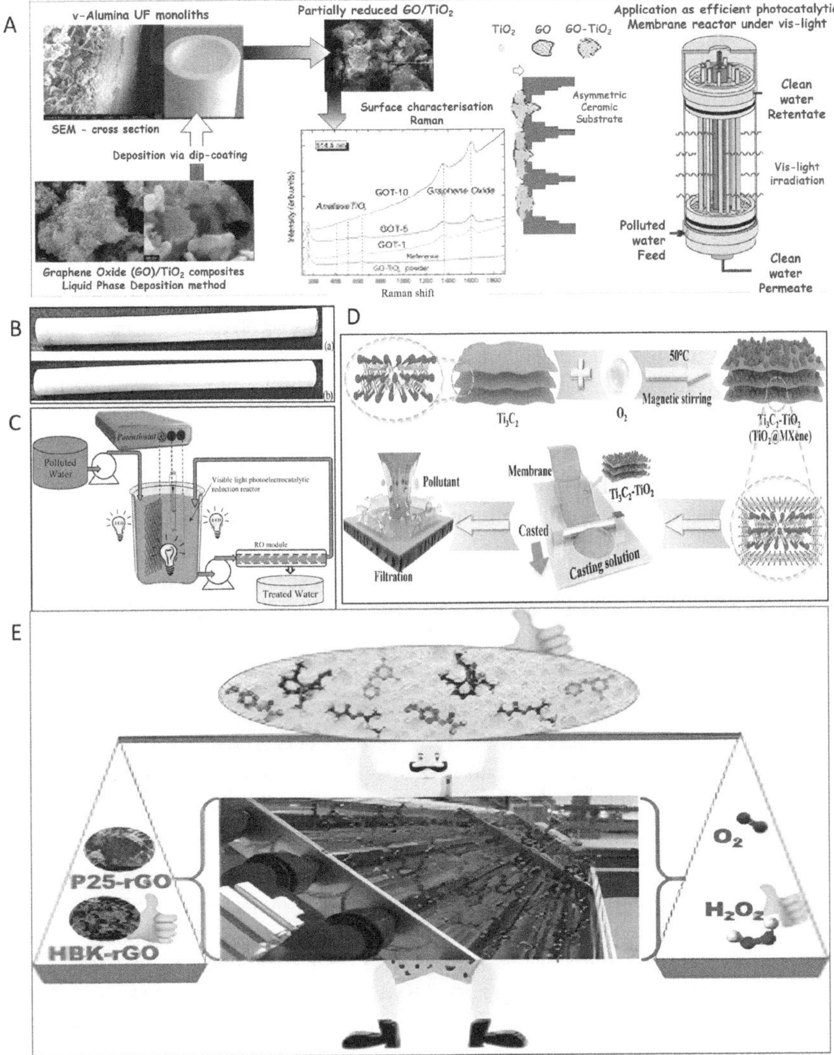

FIGURE 8.1 (A) Prototype rGO/TiO₂ composite membranes for photocatalytic UF under visible light (Athanasekou et al. 2014). (B) Photos TiO₂/graphene membrane after photocatalysis/ filtration experiments (a) Filtration experiment in dark. (b) Filtration experiment under UV (Athanasekou et al. 2014), (C) Graphene/TiO₂/Ag membrane photoreactor for photoelectrocatalysis and RO (Sheydaei and Ayoubi-Feiz 2021), (D) TiO₂@MXene as composite membranes for water purification (Huang et al. 2021) and (E) TiO₂/graphene-based solar-assisted pilot plant scale for water purification (Luna-Sanguino et al. 2020).

8.2.2 Air Purification

It has long been understood that air pollution has a severe negative impact on health. Particulate matter and gaseous pollutants are the two significant air pollutants that must be controlled. It has been shown that source control and air cleansing are successful methods for reducing air pollution. Increasing ventilation may also be beneficial in reducing indoor air pollution. While source control and treatments, such as electrostatic precipitation, wet scrubbing, and filtration, can be used to remove particulate pollutants, catalysis by semiconductors is unlikely to be efficiently dealt with (Boyjoo et al. 2017). Hence, in this section, we will be focused on the catalytic removal of various gaseous pollutants such as NO_x, SO_x, CO, ozone, volatile organic compounds (VOCs), and odorous compounds.

In early 1970, during the energy and oil crisis, photocatalytic water splitting was first employed as an effective way to remedy many energy-related issues. Later, the techniques were transferred for water and air purification purposes. The techniques used to remedy air pollution include absorption, thermo-catalytic oxidation, adsorption, bio-filtration, and photochemical oxidation. Absorption requires the migration of contaminants in the liquid phase and the pollutant's solubility in water. Deactivation and frequent replacement of the catalyst, bed poisoning, and high energy consumption are the major drawbacks in the thermo-catalytic oxidation process. Bio-filtration consists of three configurations, including bio-filters, bio-scrubbers, and bio-trickling, which require high maintenance. However, photocatalytic oxidation (PCO), which can reduce and transform air pollutants into CO_2, H_2O, and non-toxic gases, can be employed for air purification (Sharma et al. 2022). Scientists are interested in the PCO technique of degradation because of its excellent capacity to adsorb and degrade pollutants. The detailed photocatalytic mechanisms for the degradation of various pollutants like VOCs, formaldehyde, toluene, benzene, trichloroethene, NO_x, SO_x, ozone, carbon monoxide, etc. have been given in several review articles (Boyjoo et al. 2017). Generally, PCO focuses on using semiconductors that can produce ROS when light with an appropriate wavelength is absorbed. When the light (UV or visible light) of sufficient energy gets absorbed by the catalyst, electron excitation from the valance band to the conduction band results in the formation of electron-hole pairs. These exciton pairs will react with oxygen-containing molecules in the system to produce ROS, which degrades the pollutants. Figure 8.2 describes various reactions that are generally involved in the three main steps of the photocatalytic air pollutant degradation process (Bui et al. 2021).

Semiconducting nanomaterials such as TiO_2, ZnO, WO_3, Fe_2O_3, $SrTiO_3$, $BiVO_4$, Bi_2WO_6, BiO_X (X = Cl, Br, I), g-C_3N_4, etc. have been identified as efficient photocatalysts for different applications. Further, to improve the efficiency of the photocatalyst, element doping (metal or nonmetal) is employed to induce impurity states and extend the light absorption range. Another strategy to improve the catalyst's efficiency is the development of binary or ternary composites with other suitable materials to form hetero-structured interfaces, which promote charge separation and enhance light absorption (Padmanabhan et al. 2021). Also, facet modulation and oxygen vacancy engineering have become widely used interfacial charge transfer enhancement methods (Padmanabhan, Jayaraj, and John 2018). Literature in photocatalytic air purification hints at the significance of TiO_2 as an efficient, economical, abundant, and nontoxic photocatalyst that can be employed in commercial applications. Despite its favorable oxidizing potential and strong photocatalytic stability, its appeal in this field is diminished by the speedy electron-hole recombination. Bandgap tuning of TiO_2 by hybridizing with 2D graphene or graphitic carbon nitride, etc. making them a better candidate for real-time applications.

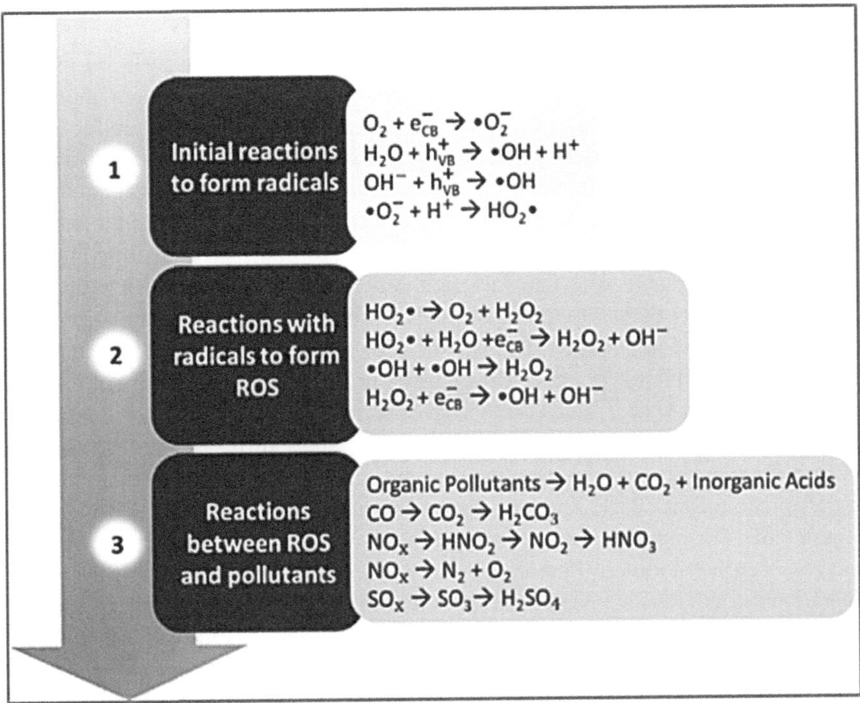

FIGURE 8.2 Various reactions involved in the catalytic degradation of air pollutants (Bui et al. 2021).

The degradation efficiency of air pollutants by the catalyst is based on several factors. If the gas flow across the photocatalyst is inadequate, the degradation rate of pollutants decreases due to inefficient diffusion of the pollutant gas molecule to the photocatalyst. Further, pollutant removal effectiveness by the catalyst reduces when the flow velocity of gases across the photocatalyst is high due to decreased contact time between ROS and pollutant. Another factor that can affect the abatement process is the light intensity used in the system. In general, higher recombination of electron-hole pairs can be observed at a high light intensity than the degradation of pollutants. The concentration of pollutants can also affect the performance of the air purifier. Similarly, the degradation rate of the gas can be affected positively and negatively by relative humidity (RH) in the reactor. When there is an overabundance of water, the reaction rate slows because the water molecule replaces the photocatalyst's reaction sites. In contrast, the absence of a water molecule causes delays in the decomposition of some gaseous compounds or prevents it entirely. As a result, an optimal RH value should be considered for forming hydroxyl radicals. Furthermore, on an industrial scale, the execution of photoreactors is constrained by photon and mass transmission restrictions, which is another key engineering challenge. Aside from that, maintaining a constant flow of light irradiance across the surface of the photocatalyst is a difficult challenge. Hence the design of the photoreactor with optimal properties must be introduced for better catalyst abatement of pollutant gases (Sharma et al. 2022).

8.2.2.1 Reactors for Catalytic Degradation of Air Pollutants

A suitable reactor must have a high photon flux efficacy and a large surface-to-volume ratio. A variety of photocatalytic reactors, including packed bed reactors, monolithic reactors, microreactors, and optical fiber reactors, have been reported so far (Li and Ma 2021). The primary comparison of different photoreactors has been reported (Li and Ma 2021) and it hints at the simple structure, recyclability, and easy operation with good stability for packed bed reactors compared to other reactors. The poor contact of the pollutant gas over the catalyst in the packed bed reactors can be rectified by using monolithic reactors, but still, they lack light efficiency. Implementing an optical fiber reactor with higher optical properties is relevant, but the large size of the reactors can hinder its practical application (Li and Ma 2021). A diagrammatic representation of different types of photoreactors has been given in Figure 8.3A. Because of the unique properties of microspace, microreactors have recently drawn a lot of attention. Because of their high lighting efficiency, quick blending, short diffusion routes, and large surface-to-volume ratio, microreactors that can scale up to photoreactors are interesting (Espíndola and Vilar 2020). An illustration of different microreactors is given in Figure 8.3B. For improved pollutant degradation efficiency, photoreactors and photocatalysts can be modified. Photoreactor design modifications will help find optimal operational conditions, while photocatalyst modifications will alter materials' physical and chemical properties. Another area that is under research is the module designs of the reactor. According to the literature, improvements in module design are strongly recommended for efficient air filtration using photocatalytic devices. The low effectiveness of the device modules is mostly due to (i) inefficient use of light irradiation, and (ii) short contact times, both of which have yet to be overcome. A simple option is to shape the support to increase the exposed surface. Many traditional devices propose to have one or more photocatalytic plates with holes for tubular UV lamps or with light sources (either UV lamps or LEDs) placed in between the plates. The only effective way to increase the irradiation area, as highlighted in the patent, is to increase the projected area of the light source on the photocatalytic material. Many patents address this issue by suggesting various catalytic carrier configurations (Figure 8.3C).

Module designs include devices with one or more photocatalytic plates with holes to place tubular UV lamps through them, light sources placed between the different plates, and W-shaped folded ceramic support set coaxially with the airflow around the tubular UV lamp. Other than the above, monolithic carriers with channels of different shapes and dimensions internally covered with a layer of photocatalytic materials have been suggested. Porous support of the photocatalytic active phase transparent to the light source, with the lamps passing through the catalytic sphere bed, and designs with flow patterns to increase the contact time and the efficiency of the devices are other examples (Muscetta and Russo 2021). Further, applying the developed nanomaterials for air purification immobilization of powdered catalysts is essential. Several factors influence the choice of support material for immobilizing photocatalysts. In general, photocatalytic activity should not be inhibited upon immobilization. Furthermore, the catalyst powders must have high adhesion properties for the support material because any nanoparticles discharged could endanger human health. Other important parameters for the supporting material include a large specific surface area, low cost, and chemical inertness. The development of passive and active ambient air

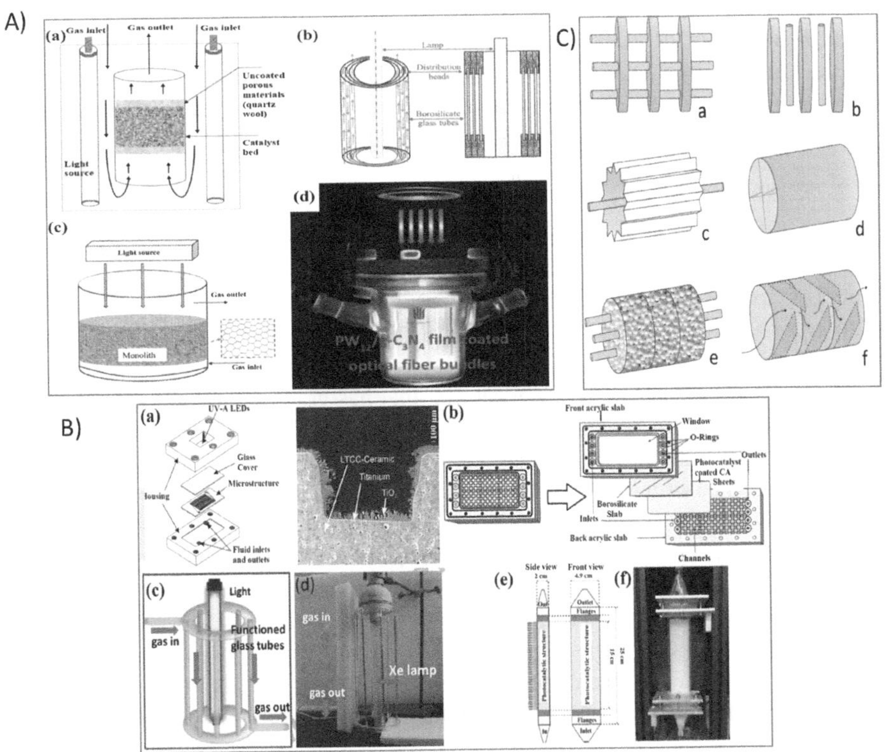

FIGURE 8.3 (A) Schematic representation (Sharma et al. 2022) of (a) packed-bed photoreactor, (b) monolithic photoreactor with multi-loop structure, (c) monolithic photoreactor with honeycomb structure, and (d) optical fiber reactor (B) Schematic representation of different microreactors (Sharma et al. 2022); (a) Micro-structured photoreactor (b) NET mix micro-reactor (c, d) Photocatalytic hybrid capillary reactor (e) and (f) Image of translucent packed-bed microstructure photoreactor. (C) Representation for different module designs suggested (Muscetta and Russo 2021): (a) Light sources passing through the catalytic support; (b) alternate light sources and catalytic support; (c) catalyst support shaped around the light source; (d) monolithic multichannel support; (e) packed bed or fibrous catalytic support; (f) baffles to enhance local turbulence.

purifying methods using the prepared air filters is relevant. The passive systems can be in the roadside, indoor environments, tunnels, and indoor parking lots. However, the active systems can be implemented in association with light sources and blowers. The active systems implemented worldwide for catalytic air purification as indoor air cleaners, catalytic street lamps, photocatalytic solar towers for an Industrial Zone, etc. (Cao, Huang, and Zhang 2021). For instance, photocatalytic concrete, which exhibits air-purifying properties, was developed, and the concrete tiles were implemented in highly trafficked canyon streets, road tunnels, etc. When exposed to sunlight, the catalyst on the material surface is activated, allowing contaminants from the environment to be degraded and transformed into less damaging compounds. It is a potential technology for reducing the number of air pollutants, particularly in polluted places

(Xu et al. 2020). Commercialized air purifiers from manufacturers, such as Sun-Pure, Diakin Industries, Ivation, Triad Aer, Vornado, BUNN, Airocide Air Purifier, Air Oasis, Eureka Fobes, Olansi, UNbeaten, etc., have been using traditional titania-based system for photocatalytic air purification. However, the application of 2D materials in air filters is currently under progress.

8.2.2.2 Laboratory Scale Development of Air Purifiers Based on 2D Materials

The photodegradation of gas-phase methanol by titania catalyst and the effect of different 2D graphene-based co-catalyst (graphene, graphene oxide, and rGO) on the degradation process was investigated by Roso et al. (2017). The electrospun polyacrylonitrile (PAN) membranes containing 60%–70% catalyst and graphene-based co-catalyst were used in the photoreactor. The photocatalyst mat has graphene/reduced graphene/graphene oxide within the fibers, leaving TiO_2 exposed on the outside and so the catalyst will be in touch with the pollutant. The rGO-based system has given the best degradation performance in terms of a higher rate of gas-phase methanol degradation than other catalysts. The number of reacted moles per gram of catalyst was significantly higher for the system based on $rGO\text{-}TiO_2$. The photoreactor used for the gas-phase pollutant degradation is given in Figure 8.4. In this air purification process, the methanol/air (dry) mixture was flushed to the reactor for a specific time for a fixed flow rate. The inside of the reactor with the as-prepared membranes was exposed to UV light for a specified time to initiate the degradation process. After the completion of the experiment, the resulting gas was analyzed with the help of GC-MS (Roso et al. 2017).

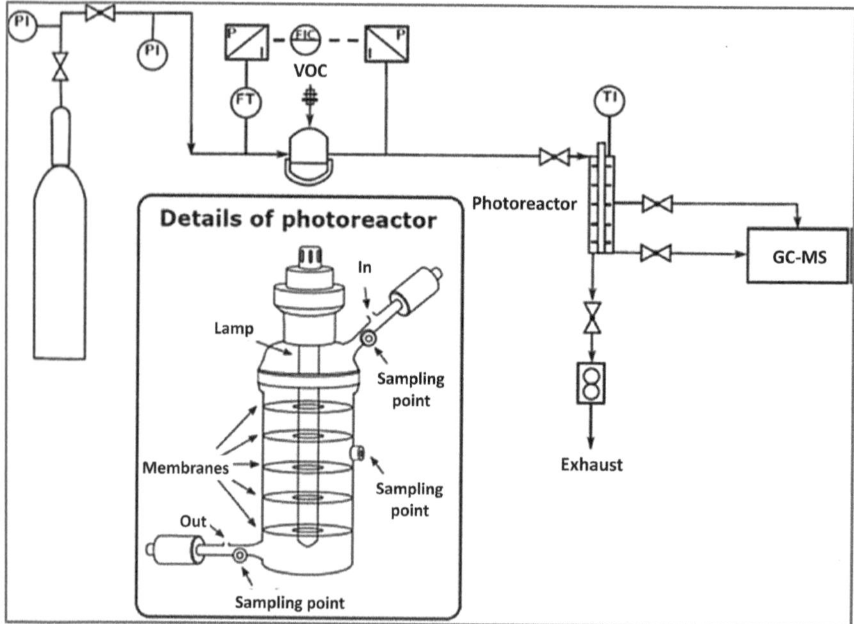

FIGURE 8.4 Photocatalytic setup diagram and detailed image of the photoreactor (Roso et al. 2017).

Later, a comprehensive study on the synthetic strategies for preparing electrospun membranes of PAN, TiO_2 graphene for air purification was reported from the same group (Figure 8.5A and B). Compared to the TiO_2-only membrane, all of the composite membranes had better photocatalytic efficiencies, which should be ascribed to the composites' synergetic charge transfer effect and a reduction in the TiO_2 bandgap. The layer-by-layer design (Type B) and the nanocomposite structure (Type C) resulted in the complete photo-degradation of gas-phase methanol to CO_2 and water, although having different kinetic constants. The degradation of VOCs and NO_2 has been effectively done by TiO_2/graphene photocatalyst (Trapalis et al. 2016). Research in developing commercializable photocatalysts based on 2D materials is in progress. For example, Tobaldi et al. established a green sol-gel synthesis approach for producing TiO_2 and TiO_2/graphene nanostructures for gas-phase purification of outdoor NO_x, benzene, and isopropanol. The photocatalytic activity of TiO_2 was doubled in NO_x abatement, benzene elimination, and isopropanol oxidation with 1.0 wt% graphene and a light mimicking the solar spectrum (Tobaldi et al. 2021). Research on developing a graphene aerogel combined with titanium dioxide for catalytic air purification has been demonstrated. TiO_2 is well known for its photocatalytic activity, while graphene aerogels are of great interest due to their exceptionally high surface area. Titania-incorporated graphene oxide aerogel was prepared with great flexibility for creating air purification cartridges (Xiong et al. 2015). Another research developed a hollow porous carbon nitride nanosphere coupled with rGO (HCNS/rGO) as a visible light photocatalyst to remove NO from the air at 600 ppb. The material with a NO removal of 64% showed excellent stability and recyclability (Wu et al. 2016). Another study used pyrolysis to create functional heterojunction foam air filters made of g-C_3N_4 and TiO_2 QDs. Under light irradiation, such 3D continuous foamy structured composite filters can provide adsorption and activation sites for oxidizing NO flow gas. The wide surface area, continuous pore channels, and excellent light-trapping capability of the material contribute to its ability as an air filter (Xiong et al. 2021). An air filter consisting of a 3D-printed flexible commercializable photocatalyst was developed by Xu et al. The composite of photocatalyst g-C_3N_4 and poly-(ethylene glycol) double acrylate with reduced bandgap showed a NO removal capability of 52.6% and good durability (Figure 8.5C) (Xu et al. 2020). An aerogel based on rGO, TiO_2, and Ag was prepared as an efficient, practically applicable photocatalyst for formaldehyde gas and liquid degradation. By combining the conventional photocatalyst (TiO_2) with Ag and rGO, a significant improvement in the transfer of photogenerated electrons was observed along with a reduction in charge recombination, and an increase in the photocatalytic reaction area. These are due to the added components' solid electrical conductivity and electron capture nature. Figure 8.5D shows the details of the rGO/TiO_2/Ag aerogel prepared for water purification applications (Wang, Wang, et al. 2019).

8.2.3 Hydrogen (H_2) Production

The energy released upon hydrogen gas combustion ~122 kJ/mol surpasses the amount of energy produced by the combustion of fossil fuels. Of total commercially produced liquid hydrogen fuel, about 48% is from natural gas, 30% is from oil, 18% is obtained from coal, and 4% is out of electrolytic water splitting.[1] Hydrogen

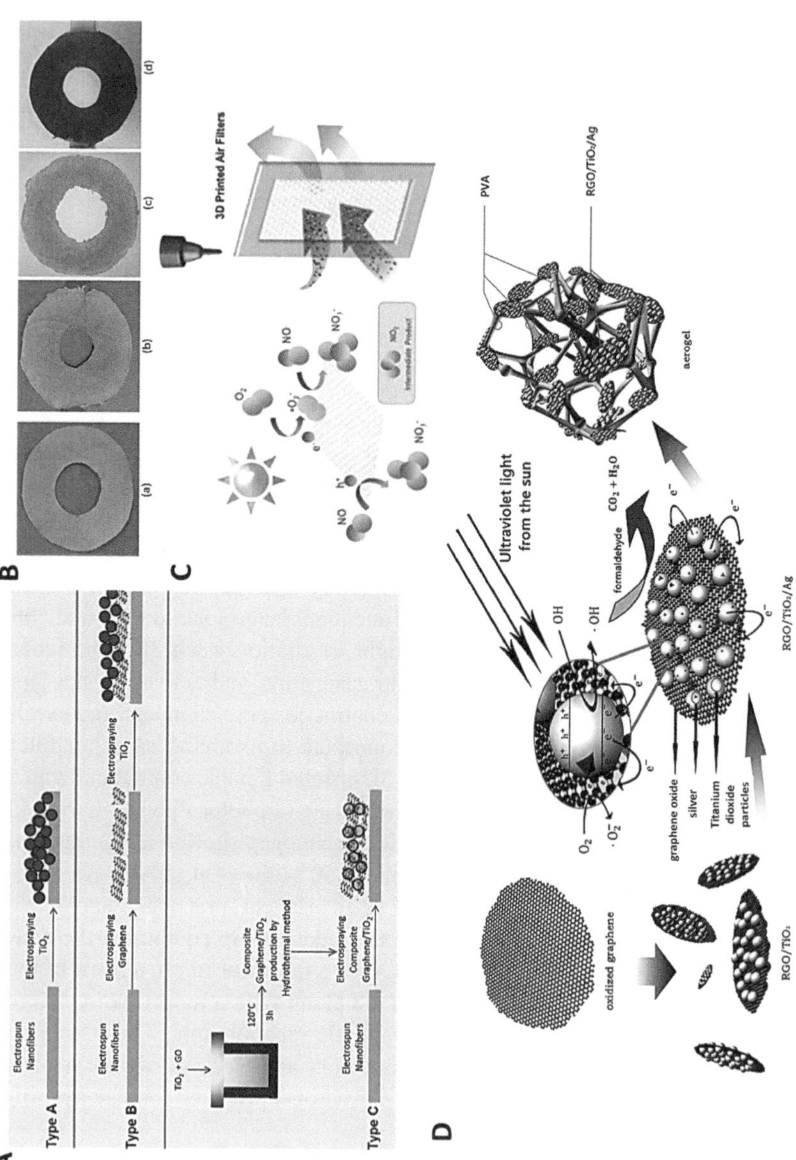

FIGURE 8.5 (A) Schematic for the development of different types of the membrane (Roso et al. 2015). (B) Photos of membrane prepared in different ways (Roso et al. 2015). (C) Schematic of the catalytic system with 3D printed air filter (Xu et al. 2020). (D) rGO/TiO$_2$/Ag aerogel prepared for water purification applications (Wang, Wang, et al. 2019).

produced from sources other than water is called blue hydrogen and via water splitting, it is called green hydrogen, this hydrogen is produced without any pollutant emission so research is mainly concentrated on the electrolytic water splitting path. Commercialization of green hydrogen is still in its infant state and many companies all over the world are trying to generate it economically. In India, five leading companies have taken prime roles – Reliance, GAIL, NTPC, Indian oil, and Larsen and Toubro (L&T). NITI Aayog, the brain of the Indian government, very recently on 29th June 2022 published an article entitled *'Harnessing Green Hydrogen-Opportunities for Deep Carbonisation in India'*. This has been viewed as a deep de-carbonization opportunity all over India. The article has put forward some suggestions and these are the following:

1. Three H_2 corridors should be built across the country based on the fuel requirement of state governments.
2. Government of each state should provide the necessary funding to researchers and industries as loans or grants to start the projects.
3. State governments should attract investors for hydrogen production as well as global level marketing of produced hydrogen.

Recently Indian Prime Minister Shri Narendra Modi announced the National Hydrogen Mission to make India the world's largest hydrogen global hub. With this move, India has kick started its journey toward green hydrogen production. Figure 8.6 shows green power supply from wind using windmills and sun via solar panels splits water in an electrolytic cell and can be safely stored in hydrogen plants.

8.2.3.1 Role of 2D Semiconductors and their Composites as Catalysts

Noble metals are very promising candidates as catalysts because of their surface plasmonic resonance phenomenon. And this makes H_2 production costlier; however,

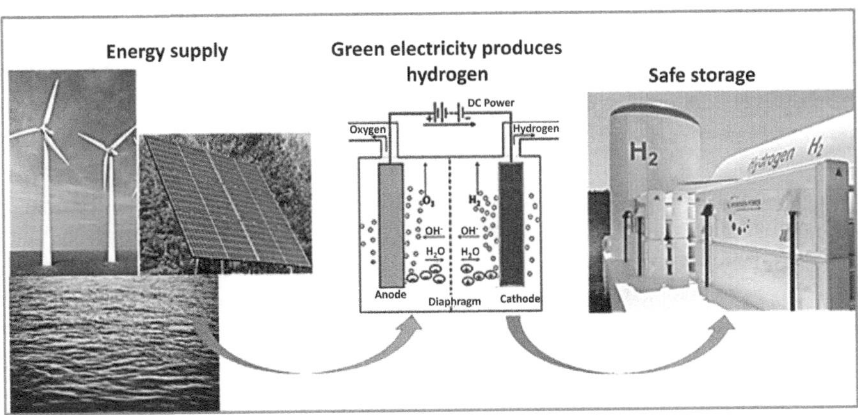

FIGURE 8.6 Wind energy and solar energy can be used as a green power supply to produce green hydrogen from water and can be safely stored in hydrogen plants before transportation to requirement spots.

manufacturing hydrogen at a large scale can reduce the cost to some extent. Also, the use of metal-free co-catalyst in addition to noble metals such as 2D semiconducting materials as an alternative can reduce cost issues. 2D materials are well known for their photocatalytic properties, because of their bandgap structure and high specific surface area. So, the merged outcome of the photochemical and electrolytic process 'photoelectrochemical (PEC)' is the better choice to increase the hydrogen evolution reaction (HER) rate. However, photochemical effects alone cannot be trusted for large-scale H_2 productions. The combined effect caused by binary complexes (metals incorporated in 2D nanomaterials) or ternary complexes (metals incorporated with more than one type of 2D nanomaterial) can further enhance HER rate. Many researchers are progressing in this direction using 2D materials such as graphene, graphene oxides, rGO, C_3N_4, boron nitride, transition metal oxides, transition metal dichalcogenides (TMDs such as MoS_2, WS_2, etc.), and layered double hydroxides (LDHs) (synthetic clays) (Faraji et al. 2019; Jones 1971). In Table 8.1, a comparative study is done for 2D materials mainly graphene, TMDs, and g-C_3N_4. 2D materials are the emerging class of photocatalysts; in the future, they may be used to increase the hydrogen production rate.

8.2.4 Carbon Dioxide (CO_2) Reduction

The emission of CO_2 greenhouse gas into the atmosphere in association with the energy crisis requires urgent attention from the scientific community. It's high time to decouple industrialization and emit pollutants into the atmosphere otherwise it leads to serious climate concerns in near future. Both commercial and residential fuel combustion contributes to CO_2 emission into the atmosphere and the rate is increasing as the population or industries are increasing. Conversion of CO_2 to various non-toxic materials such as methane, methanol, ethane, ethanol, formic acid, etc. is an ideal way to control atmospheric CO_2. The product to be formed depends on the number of charge carriers (electrons/protons) transferred between the conduction band and valence band, the nature of the catalyst, electrolyte, temperature, and pH. De-carbonization or CO_2 conversion can be achieved through various technologies – photochemical, electrochemical, photo-electrochemical, and thermochemical reactions. Among these, electrochemical CO_2 conversion reactions have more commercial visibility compared to other methods. The key parameters for the evaluation of a catalyst for electrochemical CO_2 reduction are stability of the electrocatalyst, overpotential, Tafel plot, current density, and turnover frequency. Overpotential (voltage) can be referred to as the energy required to begin the reaction, so minimum overpotential shows better catalytic properties.

A large group of semiconducting 2D materials coupled with metals and metal-free materials is being explored as a catalyst for CO_2 conversion because of their suitable bandgap and high surface area with a large number of surface active sites (Shifa et al. 2019). Graphene, graphene oxides, rGO, MOFs, TMDs (MoS_2, WS_2, etc), LDHs, etc. have been demonstrated as photocatalysts for CO_2 reduction successfully. Dimensionality or sheet-like structures play an important role in enhancing catalytic properties and stability (Buhro and Colvin 2003). Materials having a long-range order only in one plane represent a sheet like structure with high conductivity, availability,

TABLE 8.1
Comparison of reported 2D materials regarding PEC water splitting and hydrogen production

	Graphene	Transition metal dichalgogenides (TMDs)	Graphitic carbon nitride (g-C_3N_4)
Advantages	• High specific surface area. • Easy availability – the earth is abundantly composed of carbon containing materials. • High conductivity of electricity.	• Direct bandgap – An active light harvesting material. • Bandgap can be tailored accordingly by combining with other 2D materials forming heterojunctions. • Co-catalyst for HER.	• Suitable bandgap Promoting water splitting in the visible light region (Yang et al. 2017). • Easy synthesis methods, high yield, variety of functionalization methods are reported. • Thermally and chemically stable light harvesting catalyst material.
Disadvantages	• Because of high conductivity, absence of semiconducting bandgap. • Compared to modified graphene or covalent functionalized one, the pristine form shows less activity. • Intrinsic uncertainty – subjected to gradual oxidation to graphene oxide.	• Difficulty in synthesizing highly crystalline and uniform sized particles. • Conductivity is very low.	• Charge recombination is at very fast rate (Elbanna, Fujitsuka, and Majima 2017). • Low light absorption at wavelength below 460 nm (Wang, Blechert, and Antonietti 2012). • The specific surface area is poor. • Pure or unmodified g-C_3N_4 shows very low catalytic activity.
Highest reported material combo for water splitting in the literature	• Ternary complex – Pt/Si along with N-doped graphene monolayers as cathodic material (Sim et al. 2013). • At 0.25 V vs RHE, Jph = −10 mA/cm^2 • N doped graphene QDs arranged on silicon nanowires as light harvesting cathodes (Sim et al. 2015). • At 0 V vs RHE, Jph = −35 mA/cm^2 • Stability of CdSe/P25 enhanced by incorporating graphene QDs for photo anodic application (Yoon et al. 2015). • At −0.6 V vs RHE, Jph = 14.16 mA/cm^2	• N+p Si protected using Al$_2$O$_3$ incorporated over MoS$_2$ as a photocathode (Fan et al. 2017). • At 0 V vs RHE, Jph = −32 mA/cm^2 • MoSxCly along with silicon as a photocathode (Ding et al. 2016). • At 0 V vs RHE, Jph = −43 mA/cm^2 • MoS$_2$ silicon-based photocathode (Chang Kwon et al. 2017). • At 0 V vs RHE, Jph = −33.1 mA/cm^2	• g-C$_3$N$_4$/CoO/Ba-TaON based 1D/2D/nanosheet photoanode (Hou et al. 2015). • At 1.23 V vs RHE, Jph = 4.57 mA/cm^2 • C$_3$N$_4$/TiO$_2$/CQDs based anodic material (Mary Rajaitha et al. 2020). • At 1.23 V vs RHE, Jph = 2.2 mA/cm^2
Challenges	• The bandgap is absent therefore hybridization has to be done. • Better theoretical understanding is required while performing modifications, doping and covalent functionalizations.	• Efficiency and stability are relatively low for PEC reactions. • Systematic studies are needed while bandgap engineering by hybridization or modification.	• Bandgap tailoring by hybridization or doping can enhance charge transfer from the valence band to the conduction band and thereby light absorption. • Morphology and size can be changed by changing synthesis methods.

easy processability, etc. which makes them attractive. Structural properties, optical, and electronic properties can directly influence the efficiency of the reaction.

8.2.5 Microbial Disinfection

Water is an elixir – a life giving force to all manifestations of living forms on earth. The multitudes do not have access to clean and safe drinking water and many die of waterborne bacterial infections. Microbial contamination in potable water poses great pathogenic risks to humans. Humans, flora, and fauna contribute to microbial contamination through fecal matter, hospitals, industry, and cattle farms. Due to the potential for contamination by pathogenic bacteria, protozoa, or viruses, microbial-mediated pollution in aquatic settings is one of the most important concerns concerning the sanitation of water bodies used for drinking water supply, recreational activities, and seafood harvesting. These activities increase the bacterial load in a water body and reap their undeniable harm. Many remediation technologies are researched and reviewed for commercial viability. Among the many promising and investigated technologies – semiconductor photocatalysis has aroused an increasing interest as a promising environmental remediation technology for water disinfection and microbial control. Major promising categories of 2D nanomaterials, i.e., graphene, graphitic carbon nitride, 2D metal oxides and metallates, metal oxyhalides and TMDs, for photocatalytic pathogen inactivation have evoked interest and are well researched stand at the verge of commercial realization. The development of a facile, cost-effective, environment-friendly (Kumar et al. 2022), and sustainable disinfection alternative is indeed the call of the hour.

The deliberate surge toward a technology that ensures a commitment to reducing environmental footprints is all prevalent, and the world of photocatalysis is not exempted. Green chemistry with a greener source of energy, then, is much explored. Photocatalysis has a bevy of matrices based on Graphene, which are the most and probably the first matrices that would be seen in the commercial form. Furthermore, the recovery/reuse of photocatalysts and their antimicrobial potential, much desired in the aftermath of a COVID catastrophe, are well-sought attributes for future commercialization. With their decreasing costs graphene materials will be placed very soon in the market, complemented well by the high and broad photocatalytic activity against both organic pollutants and microorganisms.

The unique possibility for antibacterial action by graphene was explored first by Akhavan et al. in 2009. When combined with other antimicrobial agents, such as antimicrobial chemical compounds, metal nanoparticles, and photocatalysts, G/rGO/GO may be an effective antimicrobial agent (Akhavan and Ghaderi 2009). Applications for the antibacterial capabilities of graphene-based derivatives include bandages, gauze, and slices, paper for food packaging, face masks, water filters, and next-generation biocomposites for dental and bone implants (Janczarek et al. 2021). Consequently, it is anticipated that G/GO/rGO-based materials will soon be accessible for widespread use. Many graphene compounds and matrices have been identified and reported for microbial disinfection. The superior antibacterial performance of graphene-based materials versus pure TiO_2 is attributed to a number of factors, including the following: large surface area, superior 2D support, improved capacity to adsorb microorganisms, quick electron mobility through conjugated-bonds, and a wider spectrum of visible light absorption are just a few of the benefits. It's intriguing to see recent cases of TiO_2-graphene composite materials inactivating bacteria and genes that make them resistant to antibiotics. The laser-induced graphene (LIG) filter, a microporous conductive graphene foam, generated by direct photothermal conversion of a carbon precursor by a commercial CO_2 laser cutter, is another intriguing application of graphene that has been shown (Stanford et al. 2019).

It has been demonstrated that bacteria are effectively trapped and killed by Joule-heating, which reaches a temperature of around 380 °C when a potential is applied. The use of graphene derivatives as antiviral agents to combat the pseudorabies virus, Herpes simplex virus type 1 (HSV-1), influenza A virus (H9N2), porcine epidemic diarrhea virus, and SARS-CoV-2 has also been suggested (Srivastava et al. 2020). The key technique is blocking the viral binding sites with graphene to stop them from accessing the target cells. Direct contact with negatively charged, jagged-edged GO has been shown to destroy single-layer viral capsids. The usage of graphene and its composites, such as those with metals and polymers, is associated with high aspirations for the development of electrochemical, piezoelectric, and Cluster Regulatory Interspaced Short Polindromic Repeats CRISPR as technology-based biosensors for sensitive, early, and fast diagnosis of virus infections.

Research on several material matrices, including $TiO_2/Ag_3PO_4/graphene$ (Yang et al. 2014), MTFN membrane incorporated with rGO@TiO_2@Ag nanocomposite (Abadikhah et al. 2019), and TiO_2-rGO composite (Karaolia et al. 2018), has demonstrated good effect against *E. coli*, erythromycin- and clarithro. Ideal candidates for numerous application applications are stable and reusable photocatalytic graphene-based materials with a wide application potential, notably in the field of environmental issues. Graphene and its derivatives' antibacterial properties are particularly appealing since they might be utilized to combat bacteria, especially antibiotic-resistant strains, yeast, or filamentous fungi. There are some concerns regarding the cytotoxicity toward higher animals and human cells, and the mechanism is yet not completely known. Because of this, the main goal of the follow-up study should be to enhance scientific knowledge of the biochemical or molecular mechanisms of action that were investigated concurrently. A thorough grasp of that problem could offer excellent chances to produce smart bandages or wound dressings that are resistant to certain bacteria. Additionally, graphene-based materials offer a potentially viable technique for real-time diagnosis of the pathogen population or cancer cells in every patient, enabling the creation of new target treatments. Other potential uses include long-lasting medical devices, such as body implants, and tailored biomaterials with excellent biocompatibility (teeth and bones).

Owing to the distinctive features of possessing appropriate electronic bandgap structure and high chemical and thermal stability, metal-free polymeric nanostructured carbon-based 2D stacked graphitic carbon nitride (g-C_3N_4) photocatalyst has received significant attention (Liao 2022; Murugesan, Moses, and Chinnaswamy 2019). This photocatalyst has emerged as an economical, metal-free, and light-sensitive material for microbial remediation. 2D photocatalysts have been employed for water disinfection and microbial control as they can arrest the activity of pathogenic entities, such as bacteria, viruses, and microalgae. A symbiotic approach complementing the study of underlying inactivation mechanisms of photocatalysts for microorganisms is evaluated to provide a further understanding of g-C_3N_4-based advanced disinfection processes. Recent studies have reported the use of a novel 0D/2D p-n heterojunction of Ag_2S-anchored P-doped g-C_3N_4 (Ag_2S/PCN) (Yan et al. 2022). The photocatalytic performance of these novel composites was investigated by inactivating multidrug-resistant 7.0-log *E. coli* cells within 60 min, under visible light irradiation. This enhanced activity is attributed to the elevated visible-light response and improved production and transfer of photoexcited charges. In addition, the free radicals trapping experiments results revealed that H^+ and O_2^- were predominantly responsible for the efficient inactivation activity. The formation of reactive oxidative species holds the solution to the active disinfection of microbial species and this has been further enhanced by suitable surface modification of g-C_3N_4 (Jiang et al. 2019), evidenced by remarkably improved photocatalytic disinfection efficiency. Another interesting yet competitive class of compounds for the photocatalytic degradation of contaminants is MoS_2 (Thomas et al. 2021). The main factors affecting the photocatalytic activity of MoS_2 are its crystal structure, ease of surface chemical modification, tunable bandgap in the visible area, and the quantity and thickness of MoS_2 layers. The photocatalytic degradation process breaks down dangerous pollutants into tiny molecules by transferring electrons from reactive radical species to the contaminating molecules. Instead of using a single photocatalyst like MoS_2 or TiO_2, MoS_2/TiO_2 composites exhibit improved and synergistic photocatalytic activity. Practically speaking, a clean surface is necessary for the photocatalytic system to run continuously and degrade pollutants by photocatalysis. Despite being thoroughly studied for surfaces made of TiO_2, MoS_2 has less articles despite having strong self-cleaning qualities. The adsorption ability of such matrices is also assessed since it effectively aids in the subsequent removal of pollutants by photocatalysis. Adsorption, degradation, and removal work together in a synergistic manner to give MoS_2 exceptional self-cleaning characteristics when exposed to light. Due to the many 2D edge sites and enhanced dangling bonds present in vertically oriented 2D MoS_2 layers, these layers exhibit more chemical activity than horizontally aligned 2D MoS_2. Materials based on 2D MoS_2 have been extensively researched for use in environmental cleanup. Because of its ability to absorb light, 2D MoS_2 may capture solar energy, which is advantageous from both an economic and environmental standpoint. When switching from a 3D to a 2D MoS_2 structure, the band structure and the number of layers have a significant impact on the physical and chemical properties. Additionally, the electronic properties are well dictated by the crystal phases observed in the MoS_2, brought about by morphological changes during varied synthesis pathways. The composite material of MoS_2 with TiO_2, g-C_3N_4, and the graphene family of materials has demonstrated excellent potential for contaminant degradation.

ROS produced by semiconductor photocatalysts enables water disinfection (Ambrosi, Sofer, and Pumera 2015). Disinfection results from the oxidative stress that these ROS put on the microbes. The cell membrane of microorganism degrades, enabling the photocatalyst to function internally and causing cell death. To do this, a photon with energy greater than the bandgap is used to stimulate their bandgap energy. As a result, electrons in the VB are energized and migrate to the CB. The freshly energized electrons and the holes they have left behind can interact with dissolved oxygen and water to create ROS. This feature is used to disinfect water by causing oxidative stress on bacteria upon light irradiation on MoS_2 (Saravanan et al. 2021; Liu et al. 2016). The practical uses of MoS_2 photocatalysis are constrained by the difficulty of reusing the catalyst and the scarcity of active edge locations. As a result, there is much potential for advancement in this field of study, and serious investigation is needed before 2D MoS_2-based materials may be used in large-scale industrial applications. Numerous hypotheses have been put forth to explain how bacteria degrade, including the release of K^+ and Ca^{2+} ions from the cell membrane (Krishna et al. 2005; Yan et al. 2020). The lipid peroxidation is thought to be the primary source of photocatalytic disinfection, which inhibits normal cell functions like respiration. There was a viability loss of 91.8% and 38.9% when chemically exfoliated 2D MoS_2 sheets were compared to bulk MoS2 powder in the removal of *E. coli*. MoS_2 has been reported to be capable of rapid disinfection of water when several layers of MoS_2 are vertically aligned (Pandit et al. 2016). It was found that when the hydrophobicity of positively charged MoS_2 was modified, the mechanism of disinfection could be shifted from oxidative stress caused by ROS formation to the membrane of the bacterium being depolarized.

2D-MoS_2 sheets were found to be capable of binding to the enzymes and arresting the growth of antibiotic-resistant organisms such as methicillin-resistant *S. aureus*. By tuning the surface chemistry of the 2D-MoS_2, it is believed that enhanced inhibition could be developed. The MoS_2/α-$NiMoO_4$ composite (Ray et al. 2020) was used for the elimination of multidrug-resistant *S. aureus*, with complete inactivation obtained after 150 min. using visible light. Studies have also shown that MoS_2 composite found it to be capable of preventing horizontal gene transfer, one of the main contributions to antibiotic resistance (Wang, Qi, et al. 2019; Wang et al. 2020). Because of its magnetic characteristics, MoS_2 may be efficiently combined with Fenton (iron-based) catalytic systems to reduce the poor recyclability of MoS_2-based photocatalysts (Mu et al. 2018). Despite the rapid development of advanced photocatalysts, the practical application of photocatalytic water disinfection is still under restrictions due to the low reusability of powder photocatalysts. Binary composite heterojunction photocatalyst MoS_2/TiO_2 nanotube arrays (Yan et al. 2020) have been employed, as they have shown application prospects in water disinfection due to their excellent photocatalytic disinfection effect, recyclability, and reusability.

When the world was battered down by COVID-19, the human race inadvertently sought answers from scientists and researchers, challenging them to bring their findings to ground reality. This challenge was just one among the many crises that the world was battling until then and continues to do so. Semiconductor materials such as GO in addition to their obvious and inherent electronic properties heading up as antibacterial and anti-cancer agents owing to their high conductivity, mechanical strength, surface-to-volume ratio, and 2D monolayer structure with a negatively charged surface. GO exhibits excellent hydrophilicity with enhanced bonding and wettability, as it is surface functionalized with O-containing functional groups with carbonyl moieties and other derivatives. It is found to exhibit excellent antibacterial activity owing to the oxygen present at the basal planes. The reduced form of GO, called rGO functionalized with polysulfated dendritic polyglycerol has also exhibited considerable inactivation features against several viruses. These desired anti-microbial (Ziem et al. 2017) properties of graphene-based materials (Piperno et al. 2018) are widely explored for decontamination and disinfection of SARS-CoV-2 by constructing self-sterilized air filters and anti-viral facemasks, which are necessary prevention measures against COVID-19 (Srivastava et al. 2020).

LIG, made from a carpet of porous conductive fibers, was used in the (Ye et al. 2015) development of a self-sterilized air filter (Stanford et al. 2019), was found to work effectively in the capture of microorganisms, and facilitated the decomposition of bacteria along with other molecules and microorganism. Merging, nanosized membrane technology with a green photocatalytic process, graphene-based materials could be an effective defense against COVID-19. Similar success has been with g-C_3N_4-based photocatalysts has been substantially evaluated owing to its suitable bandgap of 2.7 eV exhibiting superior inactivation of *E. coli* and similar harmful pathogens, under UV illuminance by the generation of reactive radical species. The lineage of efficient 2D materials doesn't end abruptly. It is well accomplished with materials such as MXenes, MOFs, and COFs with desirable features involving good conductivity, layered structure, large reactive surface area, and high affinity for guest materials. 2D carbides and nitrides (Anasori, Lukatskaya, and Gogotsi 2017; Kumar et al. 2022) (MXenes) surface-functionalized (O, –OH, –F, and –Cl groups) offer significantly high surface area and porosity leading to superior adsorption of guest molecules and viruses, which has been effectively used in the Schottky heterojunction with interfacial engineering, to enhance the microbial activity of MXenes, evidenced in boost charge carrier transference (Liao et al. 2020) and the fast killing of bacteria. Many composite matrices have been promoted as protective coatings on personal protective equipment (PPE) leading to a promising alternative to develop reusable PPEs to overcome the waste crisis.

Both COFs and MOFs (Li, Chen, and Pan 2021) are a captivating class of 2D materials that meet the needs in the production of antibacterial, water adsorbents, and active filters as well as an intriguingly more specialized sector such drug delivery agents with optimal efficacy. This is made possible by its adaptable structure, biocompatibility, and capacity for encapsulation. Superior adsorption and photocatalytic virucidal and bactericidal capabilities of MOFs embedded air filters have also been investigated and reported. With its capacity to effectively identify and neutralize viruses, 2D photocatalysts have also been utilized to remove a variety of gaseous contaminants from aqueous or gaseous phases. TiO_2 has been used to create morphologically diverse textiles with exceptional mechanical strength and durability (Galante et al. 2020). Longer durability for applications in the textile, medicinal, or hygiene industries is produced by the exploration of its mechanical stability and multilayer surface roughness. Soon, PPE kits or medical apparel made from a variety of textiles that are resistant to germs, proteins, blood, and viruses will be at your disposal. For masks and air filters, paints, and disinfectants, photocatalytic materials, including metal oxides (TiO_2, WO_3, CuO, etc.), metal-free photocatalysts (graphene, GO, rGO, and g-C_3N_4), and 2D semiconductor materials are being used to stop the spread of the fatal virus in the population. For the treatment of diverse viral infections, the photothermal impact of MXenes and the targeted drug-carrier ability of MOFs/COFs are eagerly expected (Marcos-Almaraz et al.

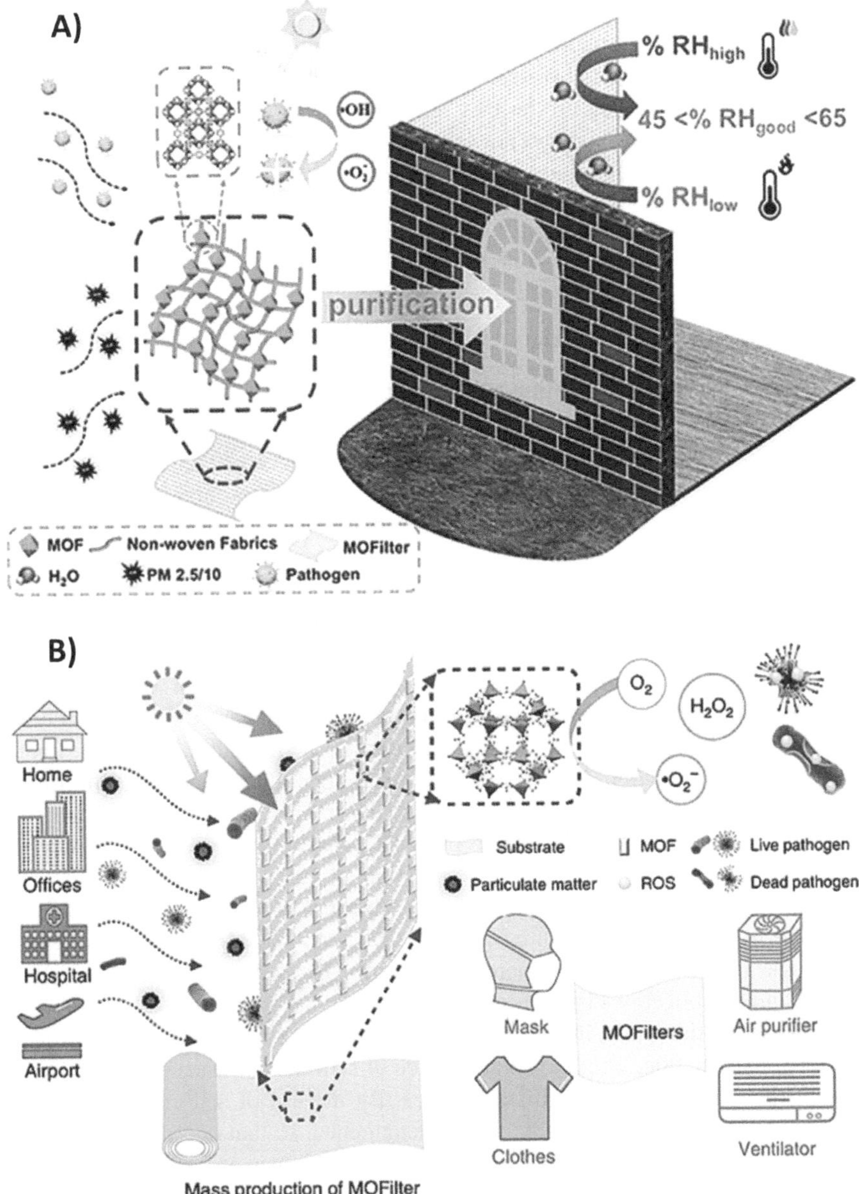

FIGURE 8.7 (A) MOF filter (CAU-1) using functionalized terephthalic acid ligands for indoor humidity, and microbial growth, to prevent pollution and (B) MOF-based filters used in different living areas (Kumar et al. 2022).

2017), (Chang et al. 2021). The MOF-based filters used for different applications are depicted in Figure 8.7.

To transition the photocatalysis process onto an industrial scale, various commercial modalities are evaluated for an effective virus photo-inactivation process.

Waking up from the forced slumber of a COVID era, current research on continuous disinfection control strategies includes unlimited usage of antiseptics and disinfectants, with their inherent corrosive nature bringing harm to lives and ecosystems. An elaborative discussion on the long-term challenges and sustainable solutions is suggested to fill in the existing knowledge gaps. Researchers need to tune down the gap between research and realization, to stand at the helm of commercializing photocatalysis processes for disinfecting virus-contaminated environments and surfaces for current and future pandemics.

8.2.6 Development in Catalytic Self-Cleaning Surfaces and Materials

8.2.6.1 Self-Cleaning Coating and Surfaces

A photocatalytic self-cleaning surface is any surface with the ability to readily remove any dirt or bacteria on it. Due to their numerous applications in a variety of fields, including interior applications in fabrics, furnishings, and window glass, as well as exterior applications in construction materials, roof tiles, car mirrors, and solar panels, photocatalytic self-cleaning materials have attracted a lot of interest in recent years. There are two different types of self-cleaning techniques: hydrophilic and hydrophobic. It has been reported that a variety of techniques, including plasma treatment, electrospinning, dip coating, layer-by-layer self-assembly, and spray coating, can create superhydrophilic self-cleaning surfaces even at large sizes for commercial applications. The relevance of introducing 2D graphene for the enhancement of the self-cleaning property of titania is evident in the literature. Under the light of a white fluorescent lamp, highly conductive and optically transparent thin films of graphene-loaded TiO_2 demonstrated dramatically improved photocatalytic activities and superhydrophilicity compared to pure TiO_2 (Anandan et al. 2013). The effective electron injection from the conduction band of TiO_2 to graphene is responsible for the increased photocatalytic activity of graphene/TiO_2 (Figure 8.8A) (Anandan et al. 2013). With longer exposure times, the graphene-loaded TiO_2 film's water contact angle increased. The commercialization of 2D semiconducting materials for self-cleaning applications is in progress.

Japan leads the world in the usage of photocatalytic self-cleaning materials in building construction. The main significance of self-cleaning surfaces is the reduction in the extent of surface dirt adherence. By the use of self-cleaning surfaces, the routine cleaning cycles can be greatly lengthened, so that one can save cleaning personnel and other related expenses. Air humidity, oxygen, and UV light are necessary for the self-cleaning process. The photocatalytic reaction can be initiated by the UV radiation present in typical daylight. Using a catalyst, usually titanium dioxide, organic dirt on a material's surface is degraded or decomposed. Japan is where self-cleaning surfaces made using photocatalysis are produced. Akira Fujishima of the University of Tokyo discovered titanium dioxide's photocatalytic ability in 1967. Photocatalytic tiles were created by the Japanese tile manufacturer Toto Ltd. and the University of Tokyo. The self-cleaning action can be achieved by activating the photocatalysis process utilizing solar irradiation as illustrated (Midtdal and Jelle 2013) in Figure 8.8B.

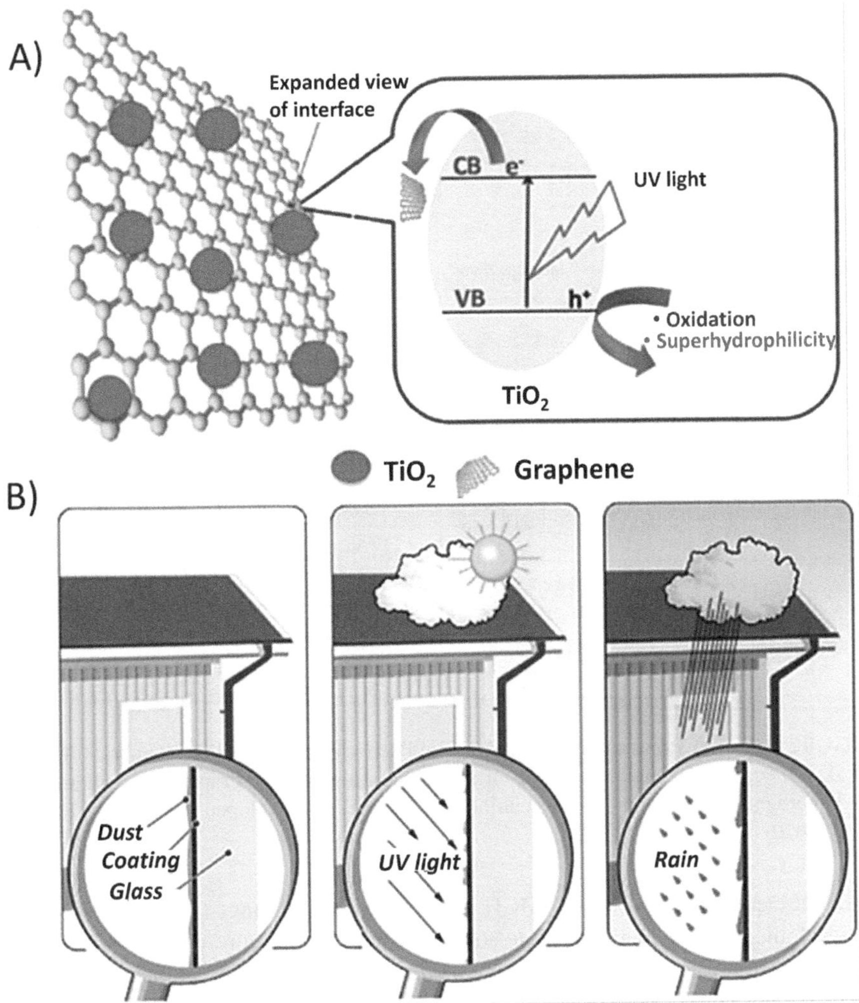

FIGURE 8.8 (A) Schematic illustration showing charge transfer in graphene/TiO$_2$ composite (Anandan et al. 2013), (B) Schematic illustration of working of self-cleaning glasses (from left to right) (Midtdal and Jelle 2013).

8.2.6.2 Self-Cleaning Membranes

In recent years, there has been a lot of interest in combining photocatalysis and membrane filtration. The implementation of a photocatalytic separation membrane with self-cleaning properties will enhance the system's usability. In a study, MCU-C$_3$N$_4$/PVDF photocatalytic membrane was developed from melamine, cyanuric acid, and urea in dimethyl sulfoxide (DMSO) as precursors. The prepared MCU(DMSO)-C$_3$N$_4$ photocatalyst material was immobilized on PVDF membranes by vacuum filtration, and polyethylene glycol and glutaraldehyde were added as crosslinkers to develop the stable membrane. They performed a four-stage filtration experiment to demonstrate

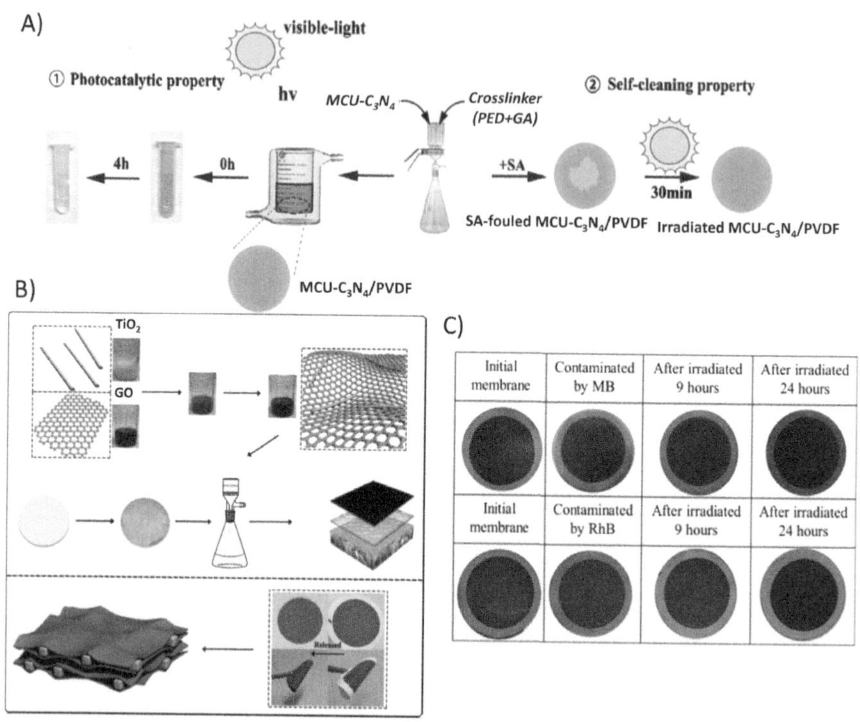

FIGURE 8.9 (A) Preparation of g-C$_3$N$_4$ based self-cleaning filtration membrane (Huang et al. 2019), (B) Schematic depiction of the preparation of TiO$_2$-graphene membranes (C) and photographs showing the self-cleaning properties of the developed membranes (Liu, Yu, et al. 2020).

the self-cleaning properties of MCU-C$_3$N$_4$/PVDF membranes (Figure 8.9A). After being fouled with a 1 wt% alginate solution, the maximal pure water flow of fouled membranes recovered to 91% of primary flux after 30 min of visible light irradiation (Huang et al. 2019). In another study, TiO$_2$ nanorod-inserted graphene-based material was prepared by vacuum filtering on a CA as a self-cleaning membrane. TiO$_2$ nanorod intercalated graphene sheets with increased interlamellar spacing showed better separation efficiency than some commercial nanofiltration membranes (Figure 8.9B). Also, it showed an extremely high rejection, particularly for dyes, for methylene blue (99.3%), rhodamine blue (99.4%), methyl orange (99.3%), and Congo red (99.6%) (Liu, Yu, et al. 2020).

8.2.6.3 Self-Cleaning Textiles

Attempts to modify traditional textile fabrics with nanomaterials to impart antibacterial activity, UV protection, fire retardancy, self-cleaning property, etc. have been carried out by various research groups. In a study, self-cleaning cotton cloth containing g-C$_3$N$_4$ catalyst was prepared by a convenient assembly method. In this, g-C$_3$N$_4$ nanosheets were anchored onto the cotton fabrics modified with PDDA through electrostatic interaction. Under simulated sunlight irradiation, the modified

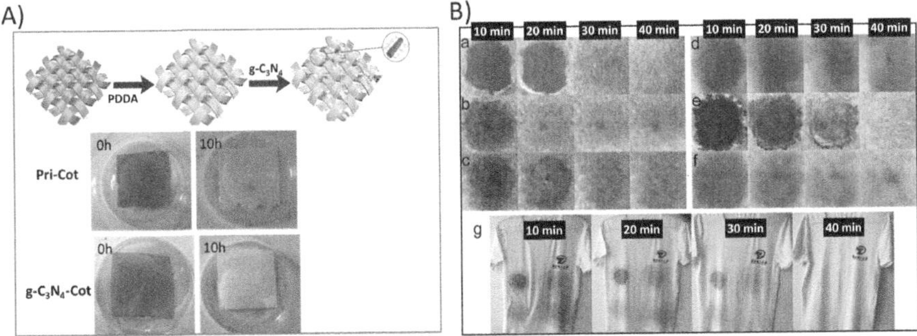

FIGURE 8.10   (A) Illustration of the process of g-C_3N_4 assembly onto cotton fabrics and the degradation of white cotton samples stained with red wine after 10 h of light (Fan et al. 2018), (B)The photocatalytic self-cleaning behavior of g-C_3N_4 modified textiles under sunlight illumination for 40 min: for different dyes ((a) RhB, (b) neutral red, (c) methyl blue, (d) reactive violet, (e) juice of red pitaya, (f) juice of waxberry), (g) the self-cleaning performance of a commercial T-shirt modified with g-C_3N_4 (Yao et al. 2019).

cotton materials demonstrated outstanding photocatalytic activity, with Rhodamine B (RhB) degradation reaching 90.2% within 80 min. Furthermore, g-C_3N_4-coated cotton fabrics demonstrated self-cleaning action for removing red wine stains under sunlight indicating their potential uses in multifunctional textiles (Figure 8.10A) (Fan et al. 2018). Another study discovered a simple method for producing 2D g-C_3N_4 nanosheet modified textiles by directly spraying g-C_3N_4 colloidal solution. The hydroxyl and amino groups on the surface of g-C_3N_4 formed due to the alkali exfoliation can form hydrogen bonds with cellulose in the textiles. Further, modifying the fabric with g-C_3N_4 did not change the textile's color and other properties. The modified fabric material showed photocatalytic self-cleaning properties for removing organic dyes (Figure 8.10B) (Yao et al. 2019).

8.3   CONCLUSIONS AND FUTURE PROSPECTS

To summarize, this chapter overviews the recent research in 2D semiconducting materials for catalytic environmental remediation. Literature shows the use of some specific systems to develop self-cleaning surfaces/coating/membranes, air filtration units, etc. This indicates the possibility of further exploration in research in catalysis using other unexplored semiconducting materials. The commercialized photocatalyst systems are mostly based on titania, while the implementation of 2D semiconducting materials in various fields of catalytic pollutant degradation or water splitting is very rare. Even though catalytic semiconductors based on 2D materials such as rGO and MXene have been reported, the underlying mechanism for the property enhancement in catalysis is unclear and yet to be explored. The mechanistic investigation of catalytic degradation of water splitting of 2D semiconductors can be done using theoretical calculations, which can help researchers to find many solutions to the experimental queries. The enhanced photocatalytic efficiency of TiO_2 after incorporating non-zero bandgap graphene-based materials such as graphite oxide, heteroatom-doped graphene, defect-rich graphene, graphene quantum dots, and nanoribbons is still under debate. This chapter gives the details of some laboratory-scale experiments based on 2D semiconducting systems, which clearly show the catalytic property enhancements when compared to the pristine conventional photocatalysts. What is lacking in the current system is the trial to implement the developed strategies into valuable products such as water or air purifiers. This may be due to the inefficiency in the large-scale synthesis of the materials. The two most difficult artificial photosynthesis reactions that produce chemical energy from solar energy are H_2 generation and CO_2 photoreduction. In the current laboratory-scale investigation, people are more into modifying the catalyst and analyzing the underlying mechanisms to develop efficient systems. Hence, commercializing these materials for wide use will take a little more time. From the engineering side, the exact design of the reactors and modules can betterment the catalytic process. Unlike in laboratory conditions, the large-scale reactors must be in accordance with the climatical, geographical, and other environmental conditions of the place under implementation. Finally, as we all are aware of the hazardous effects of nanomaterials, intensive investigation into the toxicity of different kinds of nanomaterials under study is inevitable. Hence, the various disciplines of science and technology can work together to develop sustainable and commercially viable materials for green energy production and environmental remediation.

REFERENCES

Abadikhah, Hamidreza, Ehsan Naderi Kalali, Samaneh Khodi, Xin Xu, and Simeon Agathopoulos. 2019. Multifunctional thin-film nanofiltration membrane incorporated with reduced graphene oxide@ TiO_2@ Ag nanocomposites for high desalination performance, dye retention, and antibacterial properties. *ACS Applied Materials & Interfaces* 11 (26):23535–23545.

Abdel-Maksoud, Yasmine, Emad Imam, and Adham Ramadan. 2016. TiO_2 solar photocatalytic reactor systems: selection of reactor design for scale-up and commercialization—analytical review. *Catalysts* 6 (9):138.

Akhavan, O., and E. Ghaderi. 2009. Photocatalytic reduction of graphene oxide nanosheets on TiO_2 thin film for photoinactivation of bacteria in solar light irradiation. *The Journal of Physical Chemistry C* 113 (47):20214–20220.

Ambrosi, Adriano, Zdeněk Sofer, and Martin Pumera. 2015. Lithium intercalation compound dramatically influences the electrochemical properties of exfoliated MoS_2. *Small* 11 (5):605–612.

Anandan, Srinivasan, Tata Narasinga Rao, Marappan Sathish, Dinesh Rangappa, Itaru Honma, and Masahiro Miyauchi. 2013. Superhydrophilic graphene-loaded TiO_2 thin film for self-cleaning applications. *ACS Applied Materials & Interfaces* 5 (1):207–212.

Anasori, Babak, Maria R. Lukatskaya, and Yury Gogotsi. 2017. 2D metal carbides and nitrides (MXenes) for energy storage. *Nature Reviews Materials* 2:16098.

Athanasekou, Chrysoula P., Sergio Morales-Torres, Vlassis Likodimos, et al. 2014. Prototype composite membranes of partially reduced graphene oxide/TiO_2 for photocatalytic ultrafiltration water treatment under visible light. *Applied Catalysis B: Environmental* 158–159:361–372.

Boyjoo, Yash, Hongqi Sun, Jian Liu, Vishnu K. Pareek, and Shaobin Wang. 2017. A review on photocatalysis for air treatment: from catalyst development to reactor design. *Chemical Engineering Journal* 310:537–559.

Buhro, W. E., and V. L. Colvin. 2003. Semiconductor nanocrystals: shape matters. *Nature Materials* 2 (3):138–139.

Bui, Vu Khac Hoang, Thanh Ngoc Nguyen, Vinh Van Tran, et al. 2021. Photocatalytic materials for indoor air purification systems: an updated mini-review. *Environmental Technology & Innovation* 22:101471.

Cao, Jun-ji, Yu Huang, and Qian Zhang. 2021. Ambient air purification by nanotechnologies: from theory to application. *Catalysts* 11 (11):1276.

Chang, Jian, Miao Zhang, Qiang Zhao, Liangti Qu, and Jiayin Yuan. 2021. Ultratough and ultrastrong graphene oxide hybrid films via a polycationitrile approach. *Nanoscale Horizons* 6 (4):341–347.

Chang Kwon, Ki, Seokhoon Choi, Joohee Lee, et al. 2017. Drastically enhanced hydrogen evolution activity by 2D to 3D structural transition in anion-engineered molybdenum disulfide thin films for efficient Si-based water splitting photocathodes. *Journal of Materials Chemistry A* 5 (30):15534–15542.

Choi, Myong Yong, Jayaraman Theerthagiri, and Gilberto Maia. 2021. 2D advanced materials and technologies for industrial wastewater treatment. *Chemosphere* 284:131394.

Desa, Amira Liyana, Nur Hanis Hayati Hairom, Law Yong Ng, Ching Yin Ng, Mohd Khairul Ahmad, and Abdul Wahab Mohammad. 2019. Industrial textile wastewater treatment via membrane photocatalytic reactor (MPR) in the presence of ZnO-PEG nanoparticles and tight ultrafiltration. *Journal of Water Process Engineering* 31:100872.

Ding, Qi, Bo Song, Ping Xu, and Song Jin. 2016. Efficient electrocatalytic and photoelectrochemical hydrogen generation using MoS_2 and related compounds. *Chem* 1 (5):699–726.

Elbanna, Ossama, Mamoru Fujitsuka, and Tetsuro Majima. 2017. g-C_3N_4/TiO_2 mesocrystals composite for H2 evolution under visible-light irradiation and its charge carrier dynamics. *ACS Applied Materials & Interfaces* 9 (40):34844–34854.

Enesca, Alexandru. 2021. The influence of photocatalytic reactors design and operating parameters on the wastewater organic pollutants removal—a mini-review. *Catalysts* 11 (5):556.

Espíndola, Jonathan C., and Vítor J. P. Vilar. 2020. Innovative light-driven chemical/catalytic reactors towards contaminants of emerging concern mitigation: a review. *Chemical Engineering Journal* 394:124865.

Fan, Ronglei, Jie Mao, Zhihao Yin, et al. 2017. Efficient and stable silicon photocathodes coated with vertically standing nano-MoS_2 films for solar hydrogen production. *ACS Applied Materials & Interfaces* 9 (7):6123–6129.

Fan, Yunde, Ji Zhou, Jin Zhang, et al. 2018. Photocatalysis and self-cleaning from g-C_3N_4 coated cotton fabrics under sunlight irradiation. *Chemical Physics Letters* 699:146–154.

Faraji, Monireh, Mahdieh Yousefi, Samira Yousefzadeh, et al. 2019. Two-dimensional materials in semiconductor photoelectrocatalytic systems for water splitting. *Energy & Environmental Science* 12 (1):59–95.

Fatima, Jawaria, Adnan Noor Shah, Muhammad Bilal Tahir, et al. 2022. Tunable 2D nanomaterials; their key roles and mechanisms in water purification and monitoring. *Frontiers in Environmental Science* 10:1–23.

Galante, Aj Auid-Orcid, S. Auid-Orcid Haghanifar, E. G. Romanowski, R. M. Q. Shanks, and P. W. Leu. 2020. Superhemophobic and antivirofouling coating for mechanically durable and wash-stable medical textiles. *ACS Applied Materials & Interfaces* 12 (19):22120–22128.

Han, Liping, Bo Li, Hao Wen, Yuxi Guo, and Zhan Lin. 2021. Photocatalytic degradation of mixed pollutants in aqueous wastewater using mesoporous 2D/2D TiO_2(B)-BiOBr heterojunction. *Journal of Materials Science & Technology* 70:176–184.

Hou, Jungang, Huijie Cheng, Osamu Takeda, and Hongmin Zhu. 2015. Unique 3D heterojunction photoanode design to harness charge transfer for efficient and stable photoelectrochemical water splitting. *Energy & Environmental Science* 8 (4):1348–1357.

Huang, Jinhui, Jianglin Hu, Yahui Shi, et al. 2019. Evaluation of self-cleaning and photocatalytic properties of modified g-C$_3$N$_4$ based PVDF membranes driven by visible light. *Journal of Colloid and Interface Science* 541:356–366.

Huang, Zhengyi, Qianqian Zeng, Ying Liu, et al. 2021. Facile synthesis of 2D TiO$_2$@MXene composite membrane with enhanced separation and antifouling performance. *Journal of Membrane Science* 640:119854.

Janczarek, Marcin, Maya Endo-Kimura, Zhishun Wei, et al. 2021. Novel structures and applications of graphene-based semiconductor photocatalysts: faceted particles, photonic crystals, antimicrobial and magnetic properties. *Applied Sciences* 11 (5):1982.

Jiang, Tengyao, Sijia Liu, Yangyan Gao, Asif H. Rony, Maohong Fan, and Gang Tan. 2019. Surface modification of porous g-C$_3$N$_4$ materials using a waste product for enhanced photocatalytic performance under visible light. *Green Chemistry* 21 (21):5934–5944.

Jones, Lawrence W. 1971. Liquid hydrogen as a fuel for the future. *Science* 174 (4007):367–370.

Karaolia, Popi, Irene Michael-Kordatou, Evroula Hapeshi, et al. 2018. Removal of antibiotics, antibiotic-resistant bacteria and their associated genes by graphene-based TiO$_2$ composite photocatalysts under solar radiation in urban wastewaters. *Applied Catalysis B: Environmental* 224:810–824.

Krishna, V., S. Pumprueg, S. H. Lee, et al. 2005. Photocatalytic disinfection with titanium dioxide coated multi-wall carbon nanotubes. *Process Safety and Environmental Protection* 83 (4):393–397.

Kumar, Abhinandan, Vatika Soni, Pardeep Singh, et al. 2022. Green aspects of photocatalysts during corona pandemic: a promising role for the deactivation of COVID-19 virus. *RSC Advances* 12 (22):13609–13627.

Li, Kunshan, Xinyu Lu, You Zhang, Kuiliang Liu, Yongchao Huang, and Hong Liu. 2020. Bi$_3$TaO$_7$/Ti$_3$C$_2$ heterojunctions for enhanced photocatalytic removal of water-borne contaminants. *Environmental Research* 185:109409.

Li, R. Auid-Orcid, T. Chen, and X. Pan. 2021. Metal-organic-framework-based materials for antimicrobial applications. *ACS Nano* 15 (3):3808–3848.

Li, Yu-Wei, and Wan-Li Ma. 2021. Photocatalytic oxidation technology for indoor air pollutants elimination: a review. *Chemosphere* 280:130667.

Liao, Guangfu. 2022. *Synthesis, Characterization, and Applications of Graphitic Carbon Nitride: An Emerging Carbonaceous Material*: Elsevier.

Liao, Song-Yi, Xing-Wen Huang, Qiu-Shi Rao, et al. 2020. A multifunctional MXene additive for enhancing the mechanical and electrochemical performances of the LiNi0.8Co0.1Mn0.1O2 cathode in lithium-ion batteries. *Journal of Materials Chemistry A* 8 (8):4494–4504.

Liu, Chong, Desheng Kong, Po-Chun Hsu, et al. 2016. Rapid water disinfection using vertically aligned MoS$_2$ nanofilms and visible light. *Nature Nanotechnology* 11 (12):1098–1104.

Liu, Y., L. Wang, N. Xue, P. Wang, M. Pei, and W. Guo. 2020. Ultra-highly efficient removal of methylene blue based on graphene oxide/TiO(2)/bentonite sponge. *Materials (Basel)* 13 (4):824.

Liu, Yuchuan, Zongxue Yu, Yixin Peng, Liangyan Shao, Xiuhui Li, and Haojie Zeng. 2020. A novel photocatalytic self-cleaning TiO$_2$ nanorods inserted graphene oxide-based nanofiltration membrane. *Chemical Physics Letters* 749:137424.

Loeb, Stephanie K., Pedro J. J. Alvarez, Jonathon A. Brame, et al. 2019. The technology horizon for photocatalytic water treatment: sunrise or sunset? *Environmental Science & Technology* 53 (6):2937–2947.

Luna-Sanguino, G., A. Ruíz-Delgado, A. Tolosana-Moranchel, et al. 2020. Solar photocatalytic degradation of pesticides over TiO$_2$-rGO nanocomposites at pilot plant scale. *Science of The Total Environment* 737:140286.

Marcos-Almaraz, M. T., R. Gref, V. Agostoni, et al. 2017. Towards improved HIV-microbicide activity through the co-encapsulation of NRTI drugs in biocompatible metal organic framework nanocarriers. *Journal of Materials Chemistry B* 5 (43):8563–8569.

Mary Rajaitha, P., K. Shamsa, C. Murugan, et al. 2020. Graphitic carbon nitride nanoplatelets incorporated titania based type-II heterostructure and its enhanced performance in photoelectrocatalytic water splitting. *SN Applied Sciences* 2 (4):572.

Midtdal, Krister, and Bjørn Petter Jelle. 2013. Self-cleaning glazing products: a state-of-the-art review and future research pathways. *Solar Energy Materials and Solar Cells* 109:126–141.

Mu, Dongzhao, Zhe Chen, Hongfei Shi, and Naidi Tan. 2018. Construction of flower-like $MoS_2/Fe_3O_4/rGO$ composite with enhanced photo-Fenton like catalyst performance. *RSC Advances* 8 (64):36625–36631.

Murugesan, Pramila, J. A. Moses, and Anandharamakrishnan Chinnaswamy. 2019. Photocatalytic disinfection efficiency of 2D structure graphitic carbon nitride-based nanocomposites: a review. *Journal of Materials Science* 54:12206–12235.

Muscetta, Marica, and Danilo Russo. 2021. Photocatalytic applications in wastewater and air treatment: a patent review (2010–2020). *Catalysts* 11 (7):834.

Ochoa-Gutiérrez, Karen Sofía, Erick Tabares-Aguilar, Miguel Ángel Mueses, Fiderman Machuca-Martínez, and Gianluca Li Puma. 2018. A novel prototype offset multi tubular photoreactor (OMTP) for solar photocatalytic degradation of water contaminants. *Chemical Engineering Journal* 341:628–638.

Padmanabhan, Nisha T., Madambi K. Jayaraj, and Honey John. 2018. Mechanistic insights into CTAB assisted TiO_2 crystal growth with largely exposed high energy crystal facets. *Journal of Environmental Chemical Engineering* 6 (4):5510–5519.

Padmanabhan, Nisha T., Nishanth Thomas, Jesna Louis, et al. 2021. Graphene coupled TiO_2 photocatalysts for environmental applications: A review. *Chemosphere* 271:129506.

Pandit, Subhendu, Subbaraj Karunakaran, Sunil Kumar Boda, Bikramjit Basu, and Mrinmoy De. 2016. High antibacterial activity of functionalized chemically exfoliated MoS_2. *ACS Applied Materials & Interfaces* 8 (46):31567–31573.

Phan, Duy Dũng, Frank Babick, Minh Tan Nguyen, Benno Wessely, and Michael Stintz. 2017. Modelling the influence of mass transfer on fixed-bed photocatalytic membrane reactors. *Chemical Engineering Science* 173:242–252.

Piperno, Anna, Angela Scala, Antonino Mazzaglia, et al. 2018. Cellular signaling pathways activated by functional graphene nanomaterials. *International Journal of Molecular Sciences* 19 (11):3365.

Ray, Schindra Kumar, Dipesh Dhakal, Jin Hur, and Soo Wohn Lee. 2020. Visible light driven $MoS_2/\alpha\text{-}NiMoO_4$ ultra-thin nanoneedle composite for efficient Staphylococcus aureus inactivation. *Journal of Hazardous Materials* 385:121553.

Roso, M., A. Lorenzetti, C. Boaretti, D. Hrelja, and M. Modesti. 2015. Graphene/TiO_2 based photo-catalysts on nanostructured membranes as a potential active filter media for methanol gas-phase degradation. *Applied Catalysis B: Environmental* 176–177:225–232.

Roso, Martina, Carlo Boaretti, Maria Guglielmina Pelizzo, Annamaria Lauria, Michele Modesti, and Alessandra Lorenzetti. 2017. Nanostructured photocatalysts based on different oxidized graphenes for VOCs removal. *Industrial & Engineering Chemistry Research* 56 (36):9980–9992.

Sabouni, Rana, and Hassan Gomaa. 2019. Photocatalytic degradation of pharmaceutical micro-pollutants using ZnO. *Environmental Science and Pollution Research* 26 (6):5372–5380.

Saravanan, A., P. Senthil Kumar, S. Jeevanantham, S. Karishma, and A. R. Kiruthika. 2021. Photocatalytic disinfection of micro-organisms: mechanisms and applications. *Environmental Technology & Innovation* 24:101909.

Sharma, Sarika, Rohit Kumar, Pankaj Raizada, et al. 2022. An overview on recent progress in photocatalytic air purification: metal-based and metal-free photocatalysis. *Environmental Research* 214:113995.

Sheydaei, Mohsen, and Baharak Ayoubi-Feiz. 2021. Nitrate reduction through the visible-light photoelectrocatalysis and photoelectrocatalysis/reverse osmosis processes: assessment of graphene/Ag/N-TiO2 nanocomposite. *Journal of Water Process Engineering* 39:101856.

Shifa, Tofik Ahmed, Fengmei Wang, Yang Liu, and Jun He. 2019. Heterostructures based on 2D materials: a versatile platform for efficient catalysis. *Advanced Materials* 31 (45):1804828.

Sim, Uk, Joonhee Moon, Junghyun An, et al. 2015. N-doped graphene quantum sheets on silicon nanowire photocathodes for hydrogen production. *Energy & Environmental Science* 8 (4):1329–1338.

Sim, Uk, Tae-Youl Yang, Joonhee Moon, et al. 2013. N-doped monolayer graphene catalyst on silicon photocathode for hydrogen production. *Energy & Environmental Science* 6 (12):3658–3664.

Srivastava, A. K., N. Dwivedi, C. Dhand, et al. 2020. Potential of graphene-based materials to combat COVID-19: properties, perspectives, and prospects. *Materials Today Chemistry* 18:100385.

Stanford, Michael G., John T. Li, Yuda Chen, et al. 2019. Self-sterilizing laser-induced graphene bacterial air filter. *ACS Nano* 13 (10):11912–11920.

Sundar, Kaviya Piriyah, and S. Kanmani. 2020. Progression of photocatalytic reactors and it's comparison: a review. *Chemical Engineering Research and Design* 154:135–150.

Sutisna, M. Rokhmat, E. Wibowo, R. Murniati, Khairurrijal, and M. Abdullah. 2017. Novel solar photocatalytic reactor for wastewater treatment. *IOP Conference Series: Materials Science and Engineering* 214 (1):012010.

Thomas, N., S. Mathew, K. M. Nair, et al. 2021. 2D MoS_2: structure, mechanisms, and photocatalytic applications. *Materials Today Sustainability* 13:100073.

Tobaldi, D. M., D. Dvoranová, L. Lajaunie, et al. 2021. Graphene-TiO_2 hybrids for photocatalytic aided removal of VOCs and nitrogen oxides from outdoor environment. *Chemical Engineering Journal* 405:126651.

Trapalis, A., N. Todorova, T. Giannakopoulou, et al. 2016. TiO_2/graphene composite photocatalysts for NOx removal: a comparison of surfactant-stabilized graphene and reduced graphene oxide. *Applied Catalysis B: Environmental* 180:637–647.

Wang, Haiwang, Guanqi Wang, Yukai Zhang, et al. 2019. Preparation of RGO/TiO_2/Ag aerogel and its photodegradation performance in gas phase formaldehyde. *Scientific Reports* 9 (1):16314.

Wang, Honggui, Huachen Qi, Shujun Gong, et al. 2020. Fe_3O_4 composited with MoS_2 blocks horizontal gene transfer. *Colloids and Surfaces B: Biointerfaces* 185:110569.

Wang, Honggui, Huachen Qi, Ming Zhu, et al. 2019. MoS_2 decorated nanocomposite: Fe_2O_3@MoS_2 inhibits the conjugative transfer of antibiotic resistance genes. *Ecotoxicology and Environmental Safety* 186:109781.

Wang, Xinchen, Siegfried Blechert, and Markus Antonietti. 2012. Polymeric graphitic carbon nitride for heterogeneous photocatalysis. *ACS Catalysis* 2 (8):1596–1606.

Wu, Hongxin, Dongyun Chen, Najun Li, et al. 2016. Hollow porous carbon nitride immobilized on carbonized nanofibers for highly efficient visible light photocatalytic removal of NO. *Nanoscale* 8 (23):12066–12072.

Xiong, Jun, Jun Di, Jiexiang Xia, Wenshuai Zhu, and Huaming Li. 2018. Surface defect engineering in 2D nanomaterials for photocatalysis. *Advanced Functional Materials* 28 (39):1801983.

Xiong, Mingwen, Ying Tao, Zhishu Zhao, et al. 2021. Porous g-C_3N_4/TiO_2 foam photocatalytic filter for treating NO indoor gas. *Environmental Science: Nano* 8 (6):1571–1579.

Xiong, Xicheng, Na Ji, Chunfeng Song, and Qingling Liu. 2015. Preparation functionalized graphene aerogels as air cleaner filter. *Procedia Engineering* 121:957–960.

Xu, Xi, Shuning Xiao, Habimana Jean Willy, et al. 2020. 3D-printed grids with polymeric photocatalytic system as flexible air filter. *Applied Catalysis B: Environmental* 262:118307.

Yan, Huiling, Lizhi Liu, Rong Wang, et al. 2020. Binary composite MoS2/TiO2 nanotube arrays as a recyclable and efficient photocatalyst for solar water disinfection. *Chemical Engineering Journal* 401:126052.

Yan, Wanshu, Pingping Hu, Yiguo Jiang, et al. 2022. Ag2S nanoparticles anchored on P-doped g-C$_3$N$_4$: a novel 0D/2D p-n 2 heterojunction for superior photocatalytic inactivation of 3 multidrug-resistant *E. coli. Materials Research Express* 9 (9):095001.

Yang, Xiaofei, Jieling Qin, Yan Jiang, Rong Li, Yang Li, and Hua Tang. 2014. Bifunctional TiO$_2$/Ag$_3$PO$_4$/graphene composites with superior visible light photocatalytic performance and synergistic inactivation of bacteria. *RSC Advances* 4 (36):18627–18636.

Yang, Yilong, Songcan Wang, Yongli Li, Jinshu Wang, and Lianzhou Wang. 2017. Strategies for efficient solar water splitting using carbon nitride. *Chemistry – An Asian Journal* 12 (13):1421–1434.

Yao, Chengkai, Aili Yuan, Huanhuan Zhang, et al. 2019. Facile surface modification of textiles with photocatalytic carbon nitride nanosheets and the excellent performance for self-cleaning and degradation of gaseous formaldehyde. *Journal of Colloid and Interface Science* 533:144–153.

Ye, Shiyi, Kang Shao, Zhonghua Li, et al. 2015. Antiviral activity of graphene oxide: how sharp edged structure and charge matter. *ACS Applied Materials & Interfaces* 7 (38):21571–21579.

Yoon, Ki-Yong, Hyo-Jin Ahn, Myung-Jun Kwak, Pradheep Thiyagarajan, and Ji-Hyun Jang. 2015. Graphene quantum dot-protected cadmium selenide quantum dot-sensitized photoanode for efficient photoelectrochemical cells with enhanced stability and performance. *Advanced Optical Materials* 3 (7):907–912.

Yu, Zongxue, XiaoFang Feng, Xia Min, Xiuhui Li, LiangYan Shao, and HaoJie Zeng. 2019. RGO/PDA/Bi$_{12}$O$_{17}$C$_{12}$–TiO$_2$ composite membranes based on Bi$_{12}$O$_{17}$C$_{12}$–TiO$_2$ heterojunctions with excellent photocatalytic activity for photocatalytic dyes degradation and oil–water separation. *Journal of Materials Science: Materials in Electronics* 30:18246–18258.

Zeghioud, Hichem, Nabila Khellaf, Hayet Djelal, Abdeltif Amrane, and Mohammed Bouhelassa. 2016. Photocatalytic reactors dedicated to the degradation of hazardous organic pollutants: kinetics, mechanistic aspects, and design – a review. *Chemical Engineering Communications* 203 (11):1415–1431.

Ziem, B., W. Azab, M. F. Gholami, J. P. Rabe, N. Osterrieder, and R. Haag. 2017. Size-dependent inhibition of herpesvirus cellular entry by polyvalent nanoarchitectures. *Nanoscale* 9 (11):3774–3783.

Index